World Scientific Series in Contemporary Chemical Physics – Vol. 26

TOPOLOGICAL
FOUNDATIONS OF
ELECTROMAGNETISM

World Scientific Series in Contemporary Chemical Physics

Editor-in-Chief: M. W. Evans *(AIAS, Institute of Physics, Budapest, Hungary)*

Associate Editors: S. Jeffers *(York University, Toronto)*
D. Leporini *(University of Pisa, Italy)*
J. Moscicki *(Jagellonian University, Poland)*
L. Pozhar *(The Ukrainian Academy of Sciences)*
S. Roy *(The Indian Statistical Institute)*

---

*To view the complete list of the published volumes in the series, please visit:
http://www.worldscibooks.com/series/wsccp_series.shtml

World Scientific Series in Contemporary Chemical Physics – Vol. 26

TOPOLOGICAL
FOUNDATIONS OF
ELECTROMAGNETISM

Terence W Barrett

BSEI, Vienna, Virginia, USA

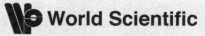 **World Scientific**

NEW JERSEY · LONDON · SINGAPORE · BEIJING · SHANGHAI · HONG KONG · TAIPEI · CHENNAI

Published by

World Scientific Publishing Co. Pte. Ltd.

5 Toh Tuck Link, Singapore 596224

USA office: 27 Warren Street, Suite 401-402, Hackensack, NJ 07601

UK office: 57 Shelton Street, Covent Garden, London WC2H 9HE

Library of Congress Cataloging-in-Publication Data
Barrett, T. W. (Terence William), 1939–
    Topological foundations of electromagnetism / Terence W. Barrett.
      p. cm. -- (World Scientific series in contemporary chemical physics ; v. 26)
    Includes bibliographical references and index.
    ISBN-13 978-981-277-996-0
    ISBN-10 981-277-996-5
    1. Electromagnetism.
    QC760 .B282 2008
    537--dc22

                                                                    2008299831

British Library Cataloguing-in-Publication Data
A catalogue record for this book is available from the British Library.

First published 2008 (Hardcover)
Reprinted 2016 (in paperback edition)
ISBN 978-981-3203-47-1

Typeset by Stallion Press
Email: enquiries@stallionpress.com

Printed in Singapore

Preface

Maxwell's equations are foundational to electromagnetic theory. They are the cornerstone of a myriad of technologies and are basic to the understanding of innumerable effects. Yet there are a few effects or phenomena that cannot be explained by the conventional Maxwell theory. This book examines those anomalous effects and shows that they can be interpreted by a Maxwell theory that is subsumed under gauge theory. Moreover, in the case of these few anomalous effects, and when Maxwell's theory finds its place in gauge theory, the conventional Maxwell theory must be extended, or generalized, to a non-Abelian form.

The tried-and-tested conventional Maxwell theory is of Abelian form. It is correctly and appropriately applied to, and explains, the great majority of cases in electromagnetism. What, then, distinguishes these cases from the aforementioned anomalous phenomena? It is the thesis of this book that it is the topology of the spatiotemporal situation that distinguishes the two classes of effects or phenomena, and the topology that is the final arbiter of the correct choice of group algebra — Abelian or non-Abelian — to use in describing an effect.

Therefore, the most basic explanation of electromagnetic phenomena and their physical models lies not in differential calculus or group theory, useful as they are, but in the topological description of the (spatiotemporal) situation. Thus, this book shows that only after the topological description is provided can understanding move to an appropriate and now-justified application of differential calculus or group theory.

Terence W. Barrett

Contents

Chapter 2: The Sagnac Effect: A Consequence of Conservation of Action Due to Gauge Field Global Conformal Invariance in a Multiply Joined Topology of Coherent Fields   95

Chapter 3: Topological Approaches to Electromagnetism   **141**

Electromagnetic Phenomena Not Explained by Maxwell's Equations[260]

Overview

The conventional Maxwell theory is a classical linear theory in which the scalar and vector potentials appear to be arbitrary and defined by boundary conditions and choice of gauge. The conventional wisdom in engineering is that potentials have only mathematical, not physical, significance. However, besides the case of quantum theory, in which it is well known that the potentials are physical constructs, there are a number of physical phenomena — *both classical and quantum-mechanical* — which indicate that the $\mathbf{A}_\mu$ fields, $\mu = 0, 1, 2, 3$, *do* possess physical significance as global-to-local operators or gauge fields, in precisely constrained topologies.

Maxwell's linear theory is of U(1) symmetry form, with Abelian commutation relations. It can be extended to include physically meaningful $\mathbf{A}_\mu$ effects by its reformulation in SU(2) and higher symmetry forms. The commutation relations of the conventional classical Maxwell theory are Abelian. When extended to SU(2) or higher symmetry forms, Maxwell's theory possesses non-Abelian commutation relations, and addresses global, i.e. nonlocal in space, as well as local phenomena with the potentials used as local-to-global operators.

An adapted Yang–Mills interpretation of *low energy fields* is applied in the following pages — an adaptation previously applied only to *high energy fields*. This adaptation is permitted by precise definition of the topological boundary conditions of those low energy electromagnetic fields. The Wu–Yang interpretation of Maxwell's theory implies the existence of magnetic monopoles and magnetic charge. As the classical fields considered here are low energy fields, *these theoretical constructs are pseudoparticle, or instanton, low energy monopoles and charges*, rather than high energy monopoles and magnetic charge (cf. Refs. 1 and 2).

Although the term "classical Maxwell theory" has a conventional meaning, this meaning actually refers to the interpretations of Maxwell's original writings by Heaviside, Fitzgerald, Lodge and Hertz. These *later interpretations of Maxwell* actually depart in a number of significant ways from Maxwell's original intention. In Maxwell's original formulation, *Faraday's electrotonic state, the A field, was central*, making this prior-to-interpretation, original Maxwell formulation compatible with Yang–Mills theory, and naturally extendable.

The *interpreted classical Maxwell theory* is, as stated, a linear theory of U(1) gauge symmetry. The mathematical dynamic entities called solitons can be either classical or quantum-mechanical, *linear or nonlinear* (cf. Refs. 3 and 4), and describe electromagnetic waves propagating through media. However, solitons are of SU(2) symmetry form.[259] In order for the conventional *interpreted classical Maxwell theory* of U(1) symmetry to describe such entities, the theory must be extended to SU(2) form.

This recent extension of soliton theory to linear equations of motion, together with the recent demonstration that the nonlinear Schrödinger equation and the Korteweg–de-Vries equation — equations of motion with soliton solutions — are reductions of the self-dual Yang–Mills equation (SDYM),[5] are pivotal in understanding the extension of Maxwell's U(1) theory to higher order symmetry forms such as SU(2). Instantons are solutions to SDYM equations which have minimum action. The use of Ward's SDYM twistor correspondence for universal integrable systems means that instantons, twistor forms, magnetic monopole constructs and soliton forms all have a pseudoparticle SU(2) correspondence.

Prolegomena A: Physical Effects Challenging a Maxwell Interpretation

A number of physical effects strongly suggest that the Maxwell field theory of electromagnetism is incomplete. Representing the influence of the independent variable, x, and the dependent variable, y, as $x \rightarrow y$, these effects address: field(s) $\rightarrow$ free electron (F $\rightarrow$ FE), field(s) $\rightarrow$ conducting electron (F $\rightarrow$ CE), field(s) $\rightarrow$ particle (F $\rightarrow$ P), wave guide $\rightarrow$ field (WG $\rightarrow$ F), conducting electron $\rightarrow$ field(s) (CE $\rightarrow$ F) and rotating frame $\rightarrow$ field(s) (RF $\rightarrow$ F) interactions. A nonexhaustive list of these experimentally observed effects, all of which involve the $\mathbf{A}_\mu$ four-potentials (vector and scalar potentials) in a physically effective role, includes:

1. *The Aharonov–Bohm and Altshuler–Aronov–Spivak effects* (F $\rightarrow$ FE and F $\rightarrow$ CE). Ehrenberg and Siday, Aharonov and Bohm, and Altshuler, Aronov and Spivak predicted experimental results by attributing physical effects to the $\mathbf{A}_\mu$ potentials. Most commentaries in classical field theory still show these potentials as mathematical conveniences without gauge invariance and with no physical significance.
2. *The topological phase effects of Berry, Aharonov, Anandan, Pancharatnam, Chiao and Wu* (WG $\rightarrow$ F and F $\rightarrow$ P). In the WG $\rightarrow$ F version, the polarization of light is changed by changing the spatial trajectory adiabatically. The Berry–Aharonov–Anandan phase has also been demonstrated at the quantum as well as the classical level. This phase effect in parameter (momentum) space is the correlate of the Aharonov–Bohm effect in metric (ordinary) space, both involving adiabatic transport.

3. *The Josephson effect* (CE → F). At both the quantum and the macrophysical level, the free energy of the barrier is defined with respect to an A_μ potential variable (phase).
4. *The quantum Hall effect* (F → CE). Gauge invariance of the A_μ vector potential, being an exact symmetry, forces the addition of a flux quantum to result in an excitation without dependence on the electron density.
5. *The De Haas–Van Alphen effect* (F → CE). The periodicity of oscillations in this effect is determined by A_μ potential dependency and gauge invariance.
6. *The Sagnac effect* (RF → F). Exhibited in the well-known and well-used ring laser gyro, this effect demonstrates that the Maxwell theory, as presently formulated, does not make explicit the constitutive relations of free space, and does not have a built-in Lorentz invariance as its field equations are independent of the metric.

The A_μ potentials have been demonstrated to be physically meaningful constructs at the quantum level (effects 1–5), at the classical level (effects 2, 3 and 6), and at a relatively long range in the case of effect 2. In the F → CE and CE → F cases (effects 1, 3–5), the effect is limited by the temperature-dependent electron coherence length with respect to the device/antenna length.

The Wu–Yang theory attempted the completion of Maxwell's theory of electromagnetism by the introduction of a nonintegrable (path-dependent) phase factor (NIP) as a physically meaningful quantity. The introduction of this construct permitted the demonstration of A_μ potential gauge invariance and gave an explanation of the Aharonov–Bohm effect. The NIP is implied by the magnetic monopole and magnetic charge theoretical constructs viewed as *pseudoparticles* or *instantons*.[a]

The recently formulated Harmuth ansatz also addresses the incompleteness of Maxwell's theory: an amended version of Maxwell's

---

[a]The term "instanton" or "pseudoparticle" is defined as the minimum action solutions of SU(2) Yang–Mills fields in Euclidean four-space R^4.[32]

equations can be used to calculate e.m. signal velocities provided that (a) a magnetic current density and (b) a magnetic monopole theoretical construct are assumed.

Formerly, treatment of the A_μ potentials as anything more than mathematical conveniences was prevented by their obvious lack of gauge invariance.[251,252] However, gauge invariance for the A_μ potentials results from situations in which fields, firstly, have a history of separate spatiotemporal conditioning and, secondly, are mapped in a many-to-one, or global-to-local, fashion (in holonomy). Such conditions are satisfied by A_μ potentials with boundary conditions, i.e. the usual empirically encountered situation. Thus, with the correct geometry and topology (i.e. with stated boundary conditions) the A_μ potentials always have physical meaning. This indicates that Maxwell's theory can be extended by the appropriate use of topological and gauge-symmetrical concepts.

The A_μ potentials are local operators mapping global spatiotemporal conditions onto the local e.m. fields. The effect of this operation is measurable as a phase change, if there is a second, comparative mapping of differentially conditioned fields in a many-to-one (global-to-local) summation. With coherent fields, the possibility of measurement (detection) after the second mapping is maximized.

The conventional Maxwell theory is incomplete due to the neglect of (1) a definition of the A_μ potentials as operators on the local intensity fields dependent on gauge, topology, geometry and global boundary conditions; and of (2) a definition of the constitutive relations between medium-independent fields and the topology of the medium.[b] Addressing these issues extends the conventional Maxwell theory to cover physical phenomena which cannot be presently explained by that theory.

---

[b]The paper by Konopinski[253] provides a notable exception to the general lack of appreciation of the central role of the A potentials in electromagnetism. Konopinski shows that the equations from which the Lorentz potentials A_ν (**A**, ϕ) arising from the sources j_n (j, ρ) are derived are $\partial_\mu^2 A_\nu = -4\pi j_\nu/c$, $\partial_\mu A_\mu = 0$, and can displace the Maxwell equations as the basis of electromagnetic theory. The Maxwell equations follow from these equations whenever the antisymmetric field tensor $F_{\mu\nu}(\mathbf{E}, \mathbf{B}) = \partial_\mu A_\nu - \partial_\nu A_\mu$ is defined.

Prolegomena B: Interpretation of Maxwell's Original Formulation

B.1. *The Faraday–Maxwell formulation*

Central to the Maxwell formulation of electromagnetism[6–16] was the Faraday concept of the *electrotonic state* (from the new Latin *tonicus*, "of tension or tone"; from the Greek *tonos*, "a stretching"). Maxwell defined this state as the "fundamental quantity in the theory of electromagnetism" on the *changes of which* (not on its absolute magnitude) the induction current depends (Ref. 8: vol. 2, p. 540). Faraday had clearly indicated the fundamental role of this state in his two circuit experiments (Ref. 17: vol. 1, series I, 60); and Maxwell endorsed its importance: "The scientific value of Faraday's conception of an electrotonic state consists in its directing the mind to lay hold of a certain quantity, on the changes of which the actual phenomena depend." (Ref. 8, vol. 2, p. 541).

The continental European views of the time concerning propagation (e.g. those of Weber, Biot, Savart and Neumann), which Maxwell opposed, were based on the concept of action-at-a-distance. In place of action-at-a-distance, Maxwell offered *a medium characterized by polarization and strain* through which radiation propagated from one local region to another local region. Furthermore, instead of force residing in the medium, Maxwell adopted another Faraday concept: force fields, or *magnetic lines of force independent of matter or magnet*. In contrast, Weber's position was that force is dependent on relative velocity and acceleration. However, Maxwell's response was not made in rebuttal of Weber's. Rather, Maxwell believed that there are two ways of looking at the subject[14]: his own and Weber's. Certainly, Helmholtz achieved a form of synthesis of the two pictures.[18]

For Maxwell, the distinction between *quantity* and *intensity* was also central. For example, magnetic intensity was represented by a line integral and referred to the magnetic polarization of the medium. Magnetic quantity was represented by a surface integral and referred to the magnetic induction in the medium. In all cases, a medium was required and *the medium was the seat of electromagnetic phenomena*. The electromagnetic and luminiferous medium were identified.

Faraday's lines of force were said to indicate the direction of minimum pressure at every point in the medium and constituted the field concept. Thus, for Maxwell, the electromagnetic field did not exist, *sui generis*, but as a state of the medium, and the mechanical cause of differences in pressure in the medium was accounted for by the hypothesis of vortices, i.e. polarization vectors. The medium was also restricted to be one in which there was only a displacement current (with no conduction currents) and in which there were no electrical or magnetic sources. Furthermore, rather than electricity *producing* a disturbance in the medium (which was W. Thomson's view), Maxwell's field theory described the presence of electricity as a disturbance, i.e. the electricity *was* the disturbance.

Maxwell conceived of electric current as a moving system with forces communicating the motion from one part of the system to another. The nature of the forces was undefined, and did not need to be defined. As for Maxwell, the forces could be eliminated from the equations of motion by Lagrangian methods for any connected system. That is, *the equations of motion were defined only locally*. Thus, Maxwell dispensed with the dynamic forces permitting propagation through the medium; only the beginning and end of the propagating process was examined and that only locally. Therefore, only the local state of the medium, or the electrotonic intensity or state, was primary, and that corresponds to Maxwell's "F" or "α_o" — or, in modern symbols, the **A** field. This is where matters stood until 1873, with the vector potential playing a pivotal *physical* role in Maxwell's theory.

B.2. *The British Maxwellians and the Maxwell–Heaviside formulation*

Subsequently, the **A** field was *banished from playing the central role in Maxwell's theory and relegated to being a mathematical (but not physical) auxiliary*. This banishment took place during the interpretation of Maxwell's theory by the Maxwellians,[12] i.e. chiefly by Heaviside, Fitzgerald, Lodge and Hertz. The "Maxwell theory" and "Maxwell's equations" we know today are really the interpretation of

Maxwell by these Maxwellians.[12,16] It was Heaviside who "murdered the **A** field" (Heaviside's description) and whose work influenced the crucial discussion which took place at the 1888 Bath meeting of the British Association (although Heaviside was not present). The "Maxwell's equations" of today are due to Heaviside's "redressing" of Maxwell's work, and should, more accurately, be known as the "Maxwell–Heaviside equations." Essentially, Heaviside took the twenty equations of Maxwell and reduced them to the four now known as "Maxwell's equations."

The British Maxwellians, Heaviside, Fitzgerald and Lodge, may have banished scalar and vector potentials from the propagation equations, but the center of concern for them remained the *dynamic state of the medium, or ether*. The banishment of the potentials can today be justified in the light of the discussion to follow in that the Maxwell theory focused on local phenomena, and the **A** field, as we shall see, addresses global connectivity. Therefore, in order for the theory to progress, it was perhaps better that the **A** field was put aside, or at least assigned an auxiliary role, *at that time*. Heaviside's comment that the electrostatic potential was a "physical inanity" was probably correct for the 19th century but, as will be shown below, the potential regained its sanity in the 20th century — starting with the work of Hermann Weyl.

But it should be emphasized that the British Maxwellians retained the focus of theory *on the medium*. Both Heaviside and Poynting agreed that the function of a wire is as a sink into which energy passes from the medium (ether) and is convected into heat. For them, wires conduct electricity with the Poynting vector pointing at right angles to the conducting wire (cf. Ref. 19, Sec. 27-5). The modern conventional view on conduction in wires is similar, but modern theory is not straightforward about where this energy goes, yet still retains Poynting's theorem. The energy flows, not through a current-carrying wire itself, but through the medium (ether) around it — or, rather, through whatever energy-storing substance a modern theorist imagines exists in the absence of the ether. Nonetheless, Heaviside was probably correct to banish scalar and vector potentials from propagation equations due to the fact that the notion of gauge invariance

(*Mass-stabinvarianz* — see below) was not yet conceived and thus not known to the Maxwellians. However, these **A** fields still remained as a repository of energy in the electrotonic state of the medium and the redressed and interpreted Maxwell theory of the British Maxwellians remained a *true dynamic* theory of electromagnetism.

B.3. *The Hertzian and current classical formulation*

But all dynamics were banished by Hertz. Hertz banished even the stresses and strains of the medium (ether) and was vigorously opposed in this by the British Maxwellians.[12] Hertz even went far beyond his mentor, Helmholtz, in his austere operational formulation. Nonetheless, the Hertz orientation finally prevailed, and the modern "Maxwell theory" is today a system of equations describing electrodynamics *which has lost its dynamical basis.*

Another significant reinterpretation of Maxwell took place in which Heaviside was involved. The 19th century battle between Heaviside and Tait concerning the use of quaternions and culminating in the victory of Heaviside and vector analysis may also be reassessed in the light of modern developments. *Without* the concepts of gauge, global (as opposed to local) fields, nonintegrable phase factors (see below) and topological connections, the use of quaternions was getting in the way of progress. That is not to say that either quaternionic algebra or the potentials were, or are, unphysical or unimportant. It is to say, rather, that the potentials could not be understood *then* with the limited theory and mathematical tools available *then*. Certainly it is *now* realized that the algebraic formulation of electromagnetism is more complicated than to be described completely even by quaternionic algebra, and certainly more complicated than to be described by simple vector analysis (cf. Ref. 20).

But to return to Maxwell's original formulation: Maxwell *did* place the **A** field center stage and *did* use quaternionic algebra to dress his theory. We know now that quaternionic algebra is described by the SU(2) group of transformations, and vector algebra by the U(1) group of transformations. As such modern propagation phenomena such as solitons are of SU(2) form, we might even view the original Maxwell formulation as more comprehensive than that offered by the British Maxwellian interpretation, and certainly more of a dynamic theory than the physically unintuitive local theory finally adopted by Hertz. That said, when it is stated, below, that the "Maxwell equations" need extension, it is really the modern Heaviside–Hertz interpretation that is meant. The *original* Maxwell theory could have been easily extended into Yang–Mills form.

Between the time of Hertz's interpretation of Maxwell's theory and the appearance of the gauge field concepts of Hermann Weyl, there appeared Whittaker's notable *mathematical* statements[21,22] that (i) the force potential can be defined in terms of both standing waves and propagating waves and (ii) any electromagnetic field e.g. dielectric displacement and magnetic force, can be expressed in terms of the derivatives of two scalar potential functions, and also be related to an inverse square law of attraction.

Whittaker[21] commenced his statement with Laplace's equation:

$$\frac{\partial^2 V}{dx^2} + \frac{\partial^2 V}{dy^2} + \frac{\partial^2 V}{dz^2} = 0, \tag{P.1}$$

which is satisfied by the potential of any distribution of matter which attracts according to Newton's law. The potential at any point (x, y, z) of any distribution of matter of mass m, situated at the point (a, b, c) which attracts according to this law, is

$$\int_0^{2\pi} f(z + ix\cos u + iy\sin u, u)du, \tag{P.2}$$

where u is a periodic argument. The most general solution to Laplace's equation using this expression is

$$V = \int_0^{2\pi} f(z + ix\cos u + iy\sin u, u)du, \tag{P.3}$$

where f is an arbitrary function of the two arguments:

$$z + ix\cos u + iy\sin u \quad \text{and} \quad u. \tag{P.4}$$

In order to express this solution as a series of harmonic terms, Whittaker[21] showed that it is only necessary to expand the function

f as a *Taylor series* with respect to the first argument $z + ix \cos u + iy \sin u$, and as a *Fourier series* with respect to the second argument, u.

Whittaker also showed that the general solution to the partial differential wave equation,

$$\frac{\partial^2 V}{dx^2} + \frac{\partial^2 V}{dy^2} + \frac{\partial^2 V}{dz^2} = \frac{k^2 \partial^2 V}{dt^2}, \tag{P.5}$$

is

$$V = \int_0^{2\pi} \int_0^{\pi} f\left(x \sin u \cos v + y \sin u \sin v + z \cos u + \frac{t}{k}, u, v\right) du\, dv, \tag{P.6}$$

where f is now an arbitrary function of the three arguments:

$$x \sin u \cos u + y \sin u \sin v + z \cos u + \frac{t}{k}, u \quad \text{and} \quad v, \tag{P.7}$$

and can be analyzed into a simple plane wave solution. Therefore, for any force varying as the inverse square of the distance, the potential of such a force satisfies *both* Laplace's equation *and* the wave equation, and can be analyzed into simple plane waves propagating with constant velocity. The *sum of these waves*, however, does not vary with time, i.e. *they are standing waves*. Therefore, the force potential can be defined in terms of both standing waves, i.e. by a global, or nonlocal, solution, and by propagating waves, i.e. by a local solution changing in time.

Furthermore, Whittaker[22] demonstrated that any electromagnetic field, e.g. dielectric displacement and magnetic force, can be expressed in terms of the derivatives of *two* scalar potential functions, F and G, satisfying

$$\frac{\partial^2 F}{dx^2} + \frac{\partial^2 F}{dy^2} + \frac{\partial^2 F}{dz^2} - \left[\frac{1}{c^2}\right] \frac{\partial^2 F}{dt^2} = 0,$$

$$\frac{\partial^2 G}{dx^2} + \frac{\partial^2 G}{dy^2} + \frac{\partial^2 G}{dz^2} - \left[\frac{1}{c^2}\right] \frac{\partial^2 G}{dt^2} = 0. \tag{P.8}$$

Thus, Whittaker's mathematical statement related the inverse square law of force to the force potential defined in terms of both standing wave (i.e. global) and propagating wave (i.e. local) solutions. The analysis also showed that the electromagnetic force fields could be defined in terms of the derivatives of two scalar potentials. This was the state of affairs prior to Weyl's introduction of gauge fields.

The landmark work of Weyl[23–26] and Yang and Mills[27] has been matched by the conception of pseudoparticle or minimum action solutions to the Yang–Mills equations,[1] i.e. instantons. Such phenomena and the appearance of gauge structure are found in simple dynamical, or classical, systems,[28] and the concept of instanton has been the focus of intense activity in recent years (cf. Refs. 2, 29–33).

The demonstration that the nonlinear Schrödinger equation and the Korteweg–de-Vries equation — equations with soliton solutions — are reductions of the self-dual Yang–Mills equations[5] with correspondences to twistor formulations[34] has provided additional evidence concerning the direction that Maxwell's theory must take. These reductions of self-dual Yang–Mills equations are known to apply to various classical systems, depending on the choice of Lie algebra associated with the self-dual fields.[35]

It is also relevant that the soliton mathematical concept need not result only from nonlinear equations. Recently, Barut[3] and Shaaarawi and Besieris[4] have demonstrated that soliton solutions are possible in the case of *linear* de-Broglie-like wave equations. Localized oscillating finite energy solutions to the massless wave equation are derived which move like massive relativistic particles with energy $E = \lambda \omega$ and momentum $\mathbf{p} = \lambda \mathbf{k}$ (λ = const.). Such soliton solutions to linear wave equations do not spread and have a finite energy field.

1. Introduction

There are a number of reasons for questioning the completeness of the conventionally interpreted Maxwell theory of electromagnetism. It is well known that there is an arbitrariness in the definition of the $\mathbf{A}$ vector and scalar potentials, which, nevertheless, have been found very useful when used in calculations with boundary conditions known.[253] The reasons for questioning completeness are due to experimental evidence (Sec. 3), theoretical (Sec. 4) and pragmatic (Sec. 5).

An examination of the Maxwell theory may begin with the well-known Maxwell equations[c]:

Coulomb's law:

$$\nabla \cdot \mathbf{D} = 4\pi\rho; \tag{1.1}$$

Maxwell's generalization of Ampere's Law:

$$\nabla \times \mathbf{H} = \left(\frac{4\pi}{c}\right)\mathbf{J} + \left(\frac{1}{c}\right)\frac{\partial \mathbf{D}}{\partial t}; \tag{1.2}$$

the postulate of an absence of free *local* magnetic poles or the differential form of Gauss' law:

$$\nabla \cdot \mathbf{B} = 0; \tag{1.3}$$

and Faraday's Law:

$$\nabla \times \mathbf{E} + \left(\frac{1}{c}\right)\frac{\partial \mathbf{B}}{\partial t} = 0. \tag{1.4}$$

The constitutive relations of the medium-independent fields to matter are well known to be

$$\mathbf{D} = \varepsilon\mathbf{E}, \tag{1.5}$$

$$\mathbf{J} = \sigma\mathbf{E}, \tag{1.6}$$

$$\mathbf{B} = \mu\mathbf{H}. \tag{1.7}$$

Because of the postulate of an absence of free local magnetic monopoles [Eq. (1.3)], the following relation is permitted:

$$\mathbf{B} = \nabla \times \mathbf{A}, \tag{1.8}$$

but the vector potential, $\mathbf{A}$, is thus always arbitrarily defined, because the gradient of some scalar function, L, can be added leaving $\mathbf{B}$ unchanged, i.e. $\mathbf{B}$ is unchanged by the gauge transformations:

$$\mathbf{A} \rightarrow \mathbf{A}' = \mathbf{A} + \nabla\Lambda; \quad \Phi \rightarrow \Phi' = \Phi - \left(\frac{1}{c}\right)\frac{\partial\Lambda}{\partial t}. \tag{1.9}$$

This arbitrary definition of the potentials means that *any* gauge chosen is arbitrary, or, an appeal must be made to boundary conditions for any choice.

---

[c]The equations are in Gaussian or cgs units: centimeter, gram and second. The Système International (SI) or mks units are: meter, kilogram and second.

Now Eq. (1.8) permits a redefinition of Eq. (1.4):

$$\nabla \times \left(\mathbf{E} + \left(\frac{1}{c} \right) \frac{\partial \mathbf{A}}{\partial t} \right) = 0 \quad \text{(Faraday's law rewritten)}, \qquad (1.10)$$

which means that the quantity in brackets is the gradient of a scalar function, Φ, and so

$$\mathbf{E} + \left(\frac{1}{c} \right) \frac{\partial \mathbf{A}}{\partial t} = -\nabla \Phi, \quad \text{or} \quad \mathbf{E} = -\partial \Phi - \left(\frac{1}{c} \right) \frac{\partial \mathbf{A}}{\partial t}, \qquad (1.11)$$

and the Maxwell equations (1.3) and (1.4) can be redefined by the use of Eqs. (1.8) and (1.11).

The Maxwell equations (1.1) and (1.2) can also be written as

$$\partial^2 \Phi + \left(\frac{1}{c} \right) \frac{\partial (\nabla \cdot \mathbf{A})}{\partial t} = -4\pi\rho, \qquad (1.12)$$

$$\nabla^2 \mathbf{A} - \left(\frac{1}{c^2} \right) \left(\frac{\partial^2 \mathbf{A}}{\partial t^2} \right) - \nabla \left(\nabla \cdot \mathbf{A} + \left(\frac{1}{c} \right) \frac{\partial \Phi}{\partial t} \right) = -\left(\frac{4\pi}{c} \right) \mathbf{J}. \quad (1.13)$$

Since the gauge conditions (1.9) are arbitrary, a set of potentials $(\mathbf{A}, \Phi)$ can be chosen so that

$$\nabla \cdot \mathbf{A} + \left(\left(\frac{1}{c} \right) \frac{\partial \Phi}{\partial t} \right) = 0. \qquad (1.14)$$

This choice is called the Lorentz condition or the Lorentz gauge. Equations (1.12) and (1.13) can then be decoupled to obtain

$$\nabla^2 \Phi + \left(\frac{1}{c^2} \right) \frac{\partial^2 \Phi}{\partial t^2} = -4\pi\rho, \qquad (1.15)$$

$$\nabla^2 \mathbf{A} - \left(\frac{1}{c^2} \right) \frac{\partial^2 \mathbf{A}}{\partial t^2} = -\left(\frac{4\pi}{c} \right) \mathbf{J}, \qquad (1.16)$$

which is useful, because the Maxwell equations are then independent of the coordinate system chosen. Nonetheless, as $\mathbf{A}$ and Φ are not gauge-invariant, the original choice of the Lorentz gauge is arbitrary — a choice which is not an inevitable consequence of the Maxwell theory — and the resultant from that choice, namely Eqs. (1.15) and (1.16), is equally arbitrary.[d]

Then again, the arbitrariness of Eq. (1.9) is useful because it permits the choice

$$\nabla \cdot \mathbf{A} = 0. \tag{1.17}$$

Equation (1.12), which is the Maxwell equation (1.1), then permits

$$\nabla^2 \Phi = -4\pi\rho, \tag{1.18}$$

which is the instantaneous Coulomb potential, and hence the condition (1.17) is called the Coulomb or transverse gauge, because the wave equation for $\mathbf{A}$ can be expressed in terms of the transverse current:

$$\nabla^2 \mathbf{A} - \left(\frac{1}{c^2}\right)\frac{\partial^2 A}{\partial t^2} = -\left(\frac{4\pi}{c}\right)\mathbf{J}_t, \tag{1.19}$$

where $\mathbf{J}_t = \mathbf{J} - \mathbf{J}_l$, with $\mathbf{J}_l$ being the longitudinal current. This is a useful thing to do when no sources are present, but, again, as $\mathbf{A}$ and Φ are not gauge-invariant, i.e. considered to have no physical meaning, the original choice of the Coulomb gauge is arbitrary, and so is the resultant from that choice, namely Eq. (1.19).

For all that, the absence of gauge invariance (physical meaning) of the $\mathbf{A}$ vector potential and the Φ scalar potential may seem a fortunate circumstance to those using the Maxwell theory to calculate predictions. These potentials have long been considered a fortunate mathematical convenience, but *just* a mathematical convenience, with no physical meaning. These constructs lack gauge invariance, a defining characteristic of physical, rather than merely mathematical, constructs. What then is meant by a gauge and gauge invariance?

---

[d]The above account applies to the Hertzian potential and the Hertz vector which are related to $\mathbf{A}$ and ϕ. However, the Hertzian vector obeys an inhomogeneous wave equation with the polarization vector as source, whereas $\mathbf{A}$ and ϕ obey their respective wave equations with electric current and charge as source. Furthermore, the Hertzian potential is a three-component potential, whereas $\mathbf{A}$ and ϕ amount to a four-potential (cf. Ref. 252, p. 254).

2. What is a Gauge?

In 1918 Weyl[23] (see also Ref. 36) treated Einstein's general theory of relativity as if the Lorentz symmetry were an example of *global* symmetry but with only *local* coordinates defineable, i.e. the general theory was considered as a *local* theory. A consequence of Weyl's theory is that the absolute magnitude or norm of a physical vector is not treated as an absolute quantity but depends on its location in space–time. This notion was called scale (*Mass-stab*) or gauge invariance.

This concept can be understood as follows. Consider a vector at position x with norm given by $f(x)$. If the coordinates are transformed, so that the vector is now at $x + dx$, the norm is $f(x + dx)$. Using the abbreviation ∂/∂^μ, $\mu = 0, 1, 2, 3$, expanding to first order, and using Einstein's summation convention,

$$f(x + dx) = f(x) + \partial_\mu f dx^\mu. \tag{2.1}$$

If a gauge change is introduced by a multiplicative scaling factor, $S(x)$, which equals unity at x, then

$$S(x + dx) = 1 + \partial_\mu S \, dx^\mu. \tag{2.2}$$

If a vector is to be constant under change of location, then

$$Sf = f + [\partial_\mu S] f dx^\mu + [\partial_\mu f] dx^\mu \tag{2.3}$$

and, on moving, the norm changes by the amount

$$[\partial_\mu + \partial_\mu S] f dx^\mu. \tag{2.4}$$

Weyl identified $\partial_\mu S$ with the electromagnetic potential $\mathbf{A}_\mu$.

However, this suggestion was rejected (by Einstein) because the natural scale for matter is the Compton wavelength, λ, and as the wave description of matter is $\lambda = \hbar/mc\lambda$ ($\hbar$ is Planck's constant and c is the speed of light), then if, as is always assumed, the wavelength is determined by the particle's mass, m, and with $\hbar$ and c constant (according to the special theory of relativity), λ cannot depend on position without violating the special theory. When made aware of this reasoning, Weyl abandoned his proposal. So the term "gauge change" originally meant "change in length," and was withdrawn from consideration for this particular metric connotation shortly after its introduction.

But the term did not die. "Gauge invariance" managed to survive in classical mechanics because, with the potentials arbitrary, Maxwell's equations for the **E**, **B**, **H** and **D** fields have a built-in symmetry and such arbitrary potentials became a useful mathematical device for simplifying many calculations in electrodynamics, as we have seen. Nevertheless, the gauge invariance in electromagnetism for the **E**, **B**, **H** and **D** fields was regarded as only an "accidental" symmetry, and the lack of gauge invariance of the electromagnetic vector and scalar potentials was interpreted as an example of the well-known arbitrariness of the concept of the potential in classical mechanics.

But this arbitrariness of the concept of the potential did not, and does not, exist in quantum mechanics. The electromagnetic vector and scalar potentials were viewed in quantum mechanics in yet another way. Upon the development of quantum mechanics, Weyl and others realized that the original gauge theory could be given a new meaning. They realized that the phase of a wave function could be a new *local* variable. Instead of a change of scale or metric, for which it was originally introduced, a gauge transformation was reinterpreted as a change in the phase of the wave function,

$$\Psi \rightarrow \Psi \exp[-ie^{\lambda}], \tag{2.5}$$

and the gauge transformation for the potential $\mathbf{A}_{\mu}$ became

$$\mathbf{A}_{\mu} \rightarrow \mathbf{A}_{\mu} - \frac{\partial \lambda}{\partial x_{\mu}}. \tag{2.6}$$

Equations (2.5) and (2.6) together ensure that the Schrödinger formulation for a single charged particle in an electromagnetic field remains invariant to phase changes because they self-cancel.[e] Thus any change in location, for that *single* charged particle, which produces a change in the phase [Eq. (2.5)] is compensated for by a corresponding change in the potential [Eq. (2.6)]. Therefore Weyl's original idea, reinterpreted, was accepted, and the potential in quantum mechanics was viewed as a *connection* which relates phases at different locations. Nevertheless, this use and interpretation did not carry

---

[e]Wignall[255] has shown that no phase change occurs for de Broglie waves under the low velocity limit of the Lorentz transformation. However, the phase of de Broglie waves is not invariant under a change of frame described by a Galilean transformation.

over into classical mechanics and a schizoid attitude has existed to this day regarding the physical meaning of the potentials in classical and quantum mechanics. In classical mechanics the potentials were, up until recently, viewed as having only an arbitrary mathematical, not physical, meaning, as they seemed to lack gauge invariance. In quantum mechanics, however, they *are* viewed as gauge-invariant and *do* possess a physical meaning. It is an aim of this book to show that in classical mechanics the potentials can also be taken to have, under special circumstances, a physical meaning, and possess gauge invariance under certain (topological) circumstances, and gauge covariance under others.

A major impetus to rethink the physical meaning of the potentials in classical mechanics came about from the experiments examined in the next section.

3.   Empirical Reasons for Questioning the Completeness of Maxwell's Theory

3.1.   *Aharonov–Bohm (AB) and Altshuler–Aronov–Spivak (AAS) effects*

Beginning in 1959 Aharonov and Bohm[40] challenged the view that the classical vector potential produces no observable physical effects by proposing two experiments. The one which is most discussed is shown in Fig. 3.1.1. A beam of monoenergetic electrons exits from the source at X and is diffracted into two beams by the two slits in the wall at Y1 and Y2. The two beams produce an interference pattern at Z which is measured. Behind the wall is a solenoid, S, the **B** field of which points out of the paper. The postulate of the absence of a free local magnetic monopole [Maxwell equation (3) above] predicts the magnetic field outside the solenoid to be zero. Before the current is turned on in the solenoid, there should be the usually expected interference patterns seen at Z. Aharonov and Bohm predicted that if the current is turned on and due to the differently directed **A** fields in paths 1 and 2 indicated by the arrows in Fig. 3.1.1, additional phase shifts should be discernible at Z. This prediction was confirmed experimentally[41–48] and the evidence has been extensively reviewed.[49–53] Aharonov and Casher[54] have extended the theoretical treatment of the AB effect to neutral particles with

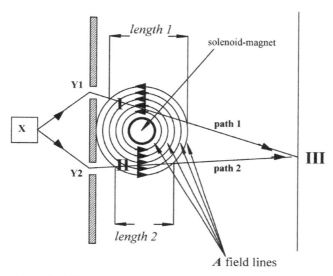

Fig. 3.1.1. Two-slit diffraction experiment of the Aharonov–Bohm effect. Electrons are produced by a source at X, diffracted by the slits at Y1 and Y2, and their diffraction pattern is detected at Z. The solenoid is between the slits and directed out of the page. The different orientations of the **A** field at the points of interaction with the two paths are indicated by the arrows > and < following the right hand rule.

a magnetic moment; and Botelho and de Mello[55] have analyzed a non-Abelian AB effect in the framework of pseudoclassical mechanics.

One explanation of the effect is as follows. Let ψ_0 be the wave function when there is no current in the solenoid. After the current is turned on, the Hamiltonian is

$$H = \frac{1}{2m}(-i\hbar\nabla\psi - e\mathbf{A})^2, \tag{3.1.1}$$

and the new wave function is

$$\psi = \psi_0 \exp\left[\frac{-ieS}{\hbar}\right], \tag{3.1.2}$$

where, **S**, the flux, is defined as

$$S = \oint \mathbf{A} \cdot d\mathbf{x}, \tag{3.1.3}$$

which is the quantum analog of the classical action evaluated along the paths 1 and 2. At point Z, the wave functions of the two electron beams are

$$\psi_1 = \psi_0 \exp\left[\frac{-eS_1}{\hbar}\right],$$

$$\psi_2 = \psi_0 \exp\left[\frac{-eS_2}{\hbar}\right],$$

(3.1.4)

and the phase difference is

$$\left(\frac{e}{\hbar}\right)(S_1 - S_2) = \left(\frac{e}{\hbar}\right)\left(\int_1 \mathbf{A} \cdot d\mathbf{x} - \int_2 \mathbf{A} \cdot d\mathbf{x}\right) = 2\pi\left(\frac{e}{\hbar}\right)\mathbf{\Phi}. \quad (3.1.5)$$

By Stokes' theorem, this is directly proportional to the magnetic flux, $\mathbf{\Phi} = \oint \mathbf{A} \cdot d\mathbf{x}$, in the solenoid.

However, the phase difference given by Eq. (3.1.5) is not single-valued. Therefore, the value of the phase change will only be determined to within an arbitrary multiple, n, of $2\pi e\mathbf{\Phi}/\hbar$, where n is the number of times the measured charge circulated the solenoid. The topological feature of the background space of the AB effect is its *multiple connectedness*.[56] Therefore, the mathematical object to be computed in this framework is a propagator expressed as a path integral in the covering space of the background physical space (cf. Ref. 57). This means that for a *simply connected* space, all paths between two points are in the same homotopy class, and the effect of the potential, A_μ, is as a multiplier of the free-particle propagator with a single gauge phase factor. *In this case, the potential has no physically discernible effect.* However, for a *multiply connected* manifold, the potential *can have a physically discernible effect* because the gauge factors can be different for different homotopy classes.[58,59]

The AB effect was confirmed experimentally in the originally proposed field $\rightarrow$ free electron (F $\rightarrow$ FE) situation (cf. Ref. 41). That is, the effect predicted by Aharonov and Bohm refers to the influence of the $\mathbf{A}$ vector potential on electrons confined to a *multiply connected region*, but within which the magnetic field is zero. As a consequence of the gauge invariance, the energy levels of the electrons have a period $\hbar/e$ of the enclosed flux. More recent experiments have addressed the appearance of the effect in the field $\rightarrow$ conduction electron (F $\rightarrow$ CE) situation (cf. Ref. 96). This situation is also not strictly the same as in the originally proposed AB experiment in another respect — the magnetic flux is produced by a large solenoid surrounding the influenced condensed matter, usually a loop or a cylinder — so that the **B** field is not set to zero within the material. However, the preponderance of the **B** field is always in the hole encompassed by that cylinder or ring, and the magnetic field causes only secondary effects in the material. In this situation, periodic oscillations in the conductance of the ring appear as a function of the applied magnetic field, **B**. The periodicity of the oscillations is

$$\nabla \mathbf{B} = \frac{\hbar}{e\mathbf{A}}, \tag{3.1.6}$$

where A is the area enclosed by the ring.

Under these conditions the AB effect is seen in normal metal,[45–47,60–67] bulk Mg,[68] semiconductors,[69,70] and on doubly connected geometries on GaAs/AlGaAs heterostructures.[71] The effect has been seen in structures such as cylindrical Mg films[72,73] and Li films,[74] wire arrays,[75,76] arrays of Ag loops,[77] small metal loops[60,65] and MBE-grown double quantum wells.[66]

Bandyopadhyay *et al.*[78] and Datta and Bandyopadhyay[70] have also discussed a novel concept for a transistor based on the electrostatic AB effect in MBE-grown quantum wells, where the current is modulated by quantum interference of electrons in two contiguous channels of a gate voltage. They predict that transistors based on this effect will have power-delay products that are orders of magnitude better than those of existing devices such as MODFETs and Josephson junctions. The transconductance will also be much higher than that of MODFETs. Unlike previous experimental treatments which assumed diffusive transport with negligible inelastic scattering, Datta and Bandyopadhyay[70] assume ballistic transport and perfect symmetry in the arms of the interferometer and in the voltage along the interferometer or two-channel structure.

Now, the AB (F → CE) effect is temperature-dependent as coherent transport is required. The effect has only been seen at very low temperatures. Measurements were made on parallel GaAs quantum wells at 4.2 K and below,[66] on 860 nm-i.d. Au loops at 0.003 K[61] and 0.05 K < T < 0.7 K,[60,63] on 75 nm-o.d. Sb loops at 0.01 < T <

1 K[79] and at 0.04 K,[64] and on Ag loop arrays at 4.2 K.[77] Measurements on 1.5–2.0-micron-diameter Mg cylinders of length 1 cm were made at 1.12 K.[72] The Thouless scaling parameter, V, or the sensitivity of energy levels to a change in the phase of the wave functions at the boundaries[80,81] implies that the necessary energy correlation range for small rings is accessible in the temperature range 0.0001–10 K.[82]

What is remarkable is that these experiments on the (F → CE) AB effect demonstrate that the effect *can* occur in *disordered* electrical conductors if the temperature is low enough. The effect in metals is a small magnetoresistance oscillation superimposed on the ohmic resistance in multiply connected conductors at low temperatures.[72,74,82] *This means that the conducting electrons must possess a high degree of phase coherence (internal correlation) over distances larger than the atomic spacing or the free path length.*[f] It was initially thought that the effects of finite temperature and the scattering from, and collision with, impurities, would cause incoherence and prevent the observation of the AB effect in bulk samples.[83] The metal loops used measure, for example, less than a micron in diameter and less than 0.1 microns in line thickness. Therefore, the electron is thought to be represented by a pair of waves — one traveling around the ring in the clockwise direction, and the other in the opposite direction, but following the time-reversed path of the first wave. Thus, although each wave has been scattered many times, *each wave collides with the same impurities*, i.e. acquires the same phase shifts, resulting in constructive interference at the origin. *The total path length of both waves is twice the circumference of the ring, meeting the requirement that the phase coherence of the electrons be larger than the circumference of the ring*, or, the transport through the metals arms considered as disordered systems is determined by the eigenvalues of a large random matrix.[83]

Thus, the conductance, G, of a one-dimensional ring in the presence of elastic scattering is[84]

$$G = \left(\frac{2e^2}{\hbar}\right)\left(\frac{t}{1-t}\right), \qquad (3.1.7)$$

---

[f]The author is indebted to an anonymous reviewer of this page for pointing out that the same coherence effect exists for electrons circulating in an antenna in the presence of ions. If this coherence effect did not exist, radiation would be impossible since emitted power would be proportional to I instead of I^2.

and an AB flux applied to the ring results in periodic oscillations of G, *provided that the phase coherence length of the ring is larger than the size of the system.* (The AB oscillations can be suppressed by magnetic fields, vanishing near resistance minima associated with plateaus in the Hall effect.[85]) Dupuis and Montambaux[86] have shown that in the case of the AB effect in metallic rings, the statistics of levels show a transition from the Gaussian orthogonal ensemble (GOE), in which the statistical ensemble shows time reversal, to the Gaussian unitary ensemble (GUE), in which time reversal symmetry is broken.

A related effect is the Altshuler–Aronov–Spivak (AAS) effect.[87] These authors considered an ultrathin normal metal cylindrical shell of moderate length but very small transverse dimensions at low temperature and how the magnetoresistance would depend on the intensity of magnetic flux axially threading the cylinder. They concluded that it would be an oscillating function of the total flux with a period of $\hbar/2e$, i.e. the same as the flux of the superconductive state. The analogous "flux quantum" of the AB effect is $\hbar/e$[60,61] and differs from the AAS situation, which involves coherent "backscattering." The AAS effect has been observed in a 1000-Å-thick magnesium layer on a quartz fiber several millimeters long.[72] More recent treatments of the AAS effect[82,88,89] are based on the quantum-mechanical transmission (t) coefficients of electrons and, unlike the original AAS treatment, find an $\hbar/e$ periodic component as well as the $\hbar/2e$ harmonic. Raising the temperature above a crossover, T_c, changes the flux periodicity of magnetic resistance oscillations from $\hbar/e$ to $\hbar/2e$, where T_c is determined by the energy correlation range $\hbar D/L^2$, where D is the elastic diffusion constant, L is the length of the sample and the quantity $\hbar D/L^2$ is the Thouless scaling parameter V for a metal.

The AAS effect arises because of a special set of trajectories — time-reversed pairs which form a closed loop — which have a fixed relative phase for any material impurity configuration. These trajectories do not average to zero and contribute to the reflection coefficients which oscillate with period $\hbar/2e$. The $\hbar/e$ oscillations of the AB effect, on the other hand, arise from oscillations in the transmission coefficients and can at higher temperatures, average to zero. Below T_c, both contributions are of order $e^2/\hbar$.[82]

Xie and DasSarma[90] studied the AB and AAS effects in the transport regime of a strongly disordered system in which electron transport is via a hopping process — specifically via variable-range-hopping transport. Their numerical results indicate that only the $\hbar/2e$ (AAS) flux-periodic oscillations survive at finite temperatures in the presence of any finite disorder.

The results of the metal loop experiments demonstrated that elastic scattering does not destroy the phase memory of the electron wave functions.[61,91] The flux periodicity in a condensed matter system due to the AB effect would not be surprising in a superconductor. However, the same periodicities in *finite* conductors is remarkable.[90] Numerical simulation of variable-range-hopping conduction[90] only finds AB oscillations ($\phi_0 = \hbar/e$) in hopping conductance when a metal ring is small and at low temperature. At the large ring limit and higher temperature AAS ($\phi_0 = \hbar/2e$) oscillations survive — a finding consistent with the experimental findings of Polyarkov *et al.*[92] A suggested reason for the retention of long range phase coherence is that the phase memory is only destroyed exponentially as $\exp[-L/L_i]$, where L_i is a "typical inelastic scattering length" and the destruction depends on the energy changes in the hopping process dependent on long wavelength, low energy acoustic phonons. The search for an explanation for both AB and AAS effects has resulted in consideration of systems as neither precisely quantum-mechanical nor classical, but in between, i.e. "mesoscopic."

"Mesoscopic" systems have been studied by Stone[93] in which the energy and spacing is only a few orders of magnitude smaller than kT at low temperatures. The prediction was made[93] that large AB oscillations should be seen in the transport coefficients of such systems. Such systems have a sample length which is much longer than the elastic mean free path, but shorter than the localization length. The magnetic field through a loop connected to leads changes the relative phase of the contribution from each arm of the loop by $2\pi\Phi/\Phi_0$, where $\Phi_0 = \hbar c/e$ is the one electron flux quantum and Φ is the flux through the hole in the loop — but only if the phase-dependent terms do not average to zero. In the mesoscopic range, if inelastic scattering is absent, these phase-dependent contributions do not self-average to zero.

Washburn *et al.*[63] and Stone and Imry[82] demonstrated experimentally that the amplitude of aperiodic and periodic conductance fluctuations decrease for the F → CE AB effect with increasing temperature. There is a characteristic correlation energy:

$$E_C = \frac{\pi \hbar D}{2L^2},$$
(3.1.8)

where D is the diffusion constant of the electrons, and L is the minimum length of the sample. If thermal energy $k_B T > E_C$, the conductance fluctuations decrease as $(E_C/k_B T)^{1/2}$. The conductance fluctuations also decrease when L_ϕ, the phase coherence length, is shorter than the length, L, or the distance between voltage probes, the decrease being described by a factor $\exp(-L/L_\phi)$.[79] This gives a conductance fluctuation:

$$\Delta Gn = \left(\frac{L_\phi}{L}\right)^{3/2}.$$
(3.1.9)

In condensed matter, therefore, the AB effect appears as the modulation of the electron wave functions by the $\mathbf{A}_\mu$ potential. The phase of the wave function can also be changed by the application of an electric field.[64] The electric field contributes to the fourth term in the four-vector product $\mathbf{A}_\mu (dx)^\mu$, which contains the scalar potential Φ associated with transverse electric fields and time. The phase shift in the wave function is

$$\Delta \phi = \int \frac{e \Phi dt}{\hbar}.$$
(3.1.10)

Experiments on Sb metal loop devices on silicon substrate have demonstrated that the voltage on capacitative probes can be used to tune the position (phase) of $\hbar/e$ oscillations in the loop. Thus, there appear to be two ways to modulate the phase of electrons in condensed matter: application of the $\mathbf{A}_\mu$ potential by threading magnetic flux between two paths of electrons; and application of a scalar potential by means of a transverse electric field. AB fluctuations in metal loops are also not symmetric about $H = 0$: four-probe measurements yield resistances which depend on the lead configurations.[62]

More recently, it has been shown that, owing to transport via edge states and penetration of a strong magnetic field into the conducting region, periodic magnetoconductance oscillations can occur in a singly connected geometry, for example as in a point contact or a "quantum dot" (a disk-shaped region in a two-dimensional electron gas).[94,95] As this effect is dependent on transport via edge states circulating along the boundary of a quantum dot and enclosing a well-defined amount of flux, the geometry is effectively doubly connected. The claim of singly-connectedness is thus more apparent than real. However, there is a difference between the AB effect in a ring and in a dot: in each period $\nabla\mathbf{B}$, the number of states below a given energy, stays constant in a ring, but increases by one in a quantum dot. In the case of a dot, the AB magnetoconductance oscillations are accompanied by an increase in the charge of the dot by one elementary charge per period. This can result in an increase in Coulomb repulsion which can block the AB magnetoconductance oscillations. This effect, occurring in quantum dots, has been called the Coulomb blockade of the AB effect.[94,95]

Finally, Boulware and Deser[97] explain the AB effect in terms of a vector potential coupling minimally to matter, i.e. a vector potential not considered as a gauge field. They provide an experimental bound on the range of such a potential as 10^2 km.

In summary, the AB and AAS effects, whether $F \rightarrow FE$ or $F \rightarrow CE$, demonstrate that the phase of a *composite* particle's wave function is a physical degree of freedom which is dependent on differences in $\mathbf{A}_\mu$ potential influences on the space–time position or path of a *first* particle's wave function with respect to that of another, *second*, particle's wave function. But the *connection*, or *mapping*, between spatiotemporally different fields or particles which originated at, or passed through, spatiotemporally separated points or paths with differential $\mathbf{A}_\mu$ potential influences, is only measurable by many-to-one mapping of those different fields or particles. By interpreting the phase of a wave function as a *local* variable instead of the norm of a vector, electromagnetism can be interpreted as a *local* gauge (phase) theory, if not exactly, then very close to the way which Weyl originally envisioned it to be.

Below, the interaction of the $\mathbf{A}_\mu$ field (x), whether vector or potential, as an independent variable with dependent variable constructs will be referred to in an $x \rightarrow y$ notation. For example, field $\rightarrow$ free electron, field → conducting electron, field → wave guide, field → neutral particle and field → rotating frame interactions will be referred to as (F → FE), (F → CE), (F → WG), (F → P) and (F → RF) interactions. Although the A_μ field is a classical field, the AB and AAS effects are either F → FE or F → CE effects and might be considered "special" in that they involve quantum-mechanical particles, i.e. electrons. In the next section, however, we examine a phase rotation which can only be considered classical, as *both* independent and dependent variables are classical. Nonetheless, the same result — the A_μ potentials demonstrate physical effects — applies.

3.2. *Topological phases: Berry, Aharonov–Anandan, Pancharatnam and Chiao–Wu phase rotation effects*

When addressing the AB effect, Wu and Yang[98] argued that the wave function of a system will be multiplied by a nonintegrable (path-dependent) phase factor after its transport around a closed curve in the presence of an A_μ potential *in ordinary space*. In the case of the Berry–Aharonov–Anandan–Pancharatnam (BAAP) phase, another nonintegrable phase factor arises from the adiabatic transport of a system around a closed path in *parameter (momentum) space*, i.e. this topological phase is the AB effect in parameter space.[99–108] The WG → F version of this effect has been experimentally verified[109] and the phase effect in general interpreted as being due to parallel transport in the presence of a gauge field.[110] The effect exists at both the classical and the quantum level (cf. Refs. 111 and 112).

There has been, however, an evolution of understanding concerning the origins of topological phase effects. Berry[99] originally proposed a geometrical (beside the usual dynamical) phase acquisition for a nondegenerate quantum state which varies adiabatically through a circuit in parameter space. Later, the constraint of an adiabatic approximation was removed[103] and also the constraint of degenerate states.[28] Then Aharonov and Anandan[113] showed that the effect can be defined for any cyclic evolution of a quantum system. Bhandari and Samuel[114] have also pointed out that Berry's phase is closely connected with a phase discovered by Pancharatnam.[115] These authors also demonstrated that unitary time evolution is not essential for the appearance of a phase change in one beam of an interferometer. This is because the polarization state of light can be taken along a closed circuit on the Poincaré sphere, and the resulting polarization changes are not necessarily a function of a unitary time evolution. Thus, the current thinking is that the history of "windings" of a particle is "remembered," or registered and indicated, by changes in phase either in a quantum-mechanical particle's state or in a classical wave's polarization. The topological phase effect appears to arise from the nontrivial topology of the complex projective Hilbert space — whether classical or quantum-mechanical[116] — and to be equivalent to a gauge potential in the parameter space of the system — again, whether classical or quantum-mechanical.

Jiao *et al.*[117] have also indicated at least three variations of topological phases: (i) the phase which arises from a cycling in the direction of a beam of light so that the tip of the spin vector of a photon in this beam traces out a closed curve on the sphere of spin directions — which is that originally studied by Chiao and Wu[106]; (ii) Pancharatnam's phase arising from a cycling in the polarization states of the light while keeping the direction of the beam of light fixed, so that Stokes' vector traces out a closed curve on the Poincaré sphere, i.e. the phase change is due to a polarization change; (iii) the phase change due to a cycle of changes in squeezed states of light.

Topological phase change effects, in the field $\rightarrow$ photon version, have been observed in NMR interferometry experiments[118,119] and using ultracold neutrons[120]; in coherent states[121,122]; optical resonance[123]; and the degenerate parametric amplifier.[124] However, topological phase change effects are more commonly studied in a classical wave guide $\rightarrow$ classical field (WG $\rightarrow$ F) version, in which the parameter space is the momentum $\mathbf{k}$ space.[106,117,125]

For example, the helicity or polarization state, σ, is[106]

$$\sigma = \mathbf{s} \cdot \mathbf{k}, \tag{3.2.1}$$

where $\mathbf{s}$ is a spin or helicity operator and $\mathbf{k}$ is the direction of propagation (k_x, k_y, k_z). If τ is the optical path length, then $|\mathbf{k}(\tau), \sigma >$ is the spin or polarization state. Interpreted classically, the constraint of keeping $\mathbf{k}$ parallel to the axis of a wave guide is due to the linear momentum being in that direction. This means that a wave guide can act as a polarization rotator. Furthermore, as helicity (polarization), σ, is adiabatically conserved, s is also constrained to remain parallel to the local axis of the wave guide. Therefore, the topology of a wave guide, e.g. a helix shape, will constrain $\mathbf{k}$, and also s, to perform a trajectory C on the surface of a sphere in the parameter space (k_x, k_y, k_z) which prescribes the linear momentum. Thus, the topology of the constrained trajectory of radiation progressing between two *local* positions has a *global* effect indicated by a polarization (spin) change. If $\gamma(C)$ is the topological phase, and $\beta = \exp[i\gamma(C)]$ is a phase factor, the final polarization state after progression along a constrained trajectory, i.e. "momentum conditioning," is

$$\sigma_2 = \beta \cdot \mathbf{s} \cdot \mathbf{k} \qquad (3.2.2)$$

where the subscript indicates a second location on the trajectory.

As a monopole is theoretically required at $\mathbf{k} = 0$, owing to the radial symmetry of the parameter space and resulting singularity, a solid angle $\Omega(C)$ can be defined on a parameter space sphere with respect to the origin $\mathbf{k} = 0$. Thus, $\Omega(C)$ can be said to define the "excited states" of the monopole at $\mathbf{k} = 0$. Therefore

$$\sigma_2 - \sigma_1 = \beta \cdot \mathbf{s} \cdot \mathbf{k} - \sigma_1 = \sigma_1 \Omega(C) - \sigma_1 = \gamma(C). \qquad (3.2.3)$$

The following question can be asked: What conservation law underlies the topological phase? A clue is provided by Kitano *et al.*,[126] who point out that the phase change can also be seen in discrete optical systems which contain no wave guides, for example in a configuration of (ideal or infinitely conducting) mirrors. Now, mirrors do not conserve helicity; they reverse it and the local tangent vector, $\mathbf{t}$, must be replaced by $-\mathbf{t}$ on alternate segments of the light path. Mirror configurations of this type have been used in a laser gyro.[127] This suggests that changes of acceleration, whether along a wave guide or in a mirror reflection, under equivalence principle conditions are the compensatory changes which match changes in the topological phase, giving the conservation equation:

$$\gamma(C) + \oint \mathbf{A} \cdot dl = 0. \qquad (3.2.4)$$

That the phase effect change can occur in classical mechanical form is witnessed by changes in polarization rotation resulting from changes in the topological path of a light beam. Tomita and Chiao[109] demonstrated effective optical activity of a helically wound single mode optical fiber in confirmation of Berry's prediction. The angle of rotation of linearly polarized light in the fiber gives a direct measure of the topological phase *at the classical level.* (Hannay[128] has also discussed the classical limit of the topological phase in the case of a symmetric top.) The effect arises from the overall geometry of the path taken by the light and is thus a *global topological effect* independent of the material properties of the fiber. The optical rotation is *independent of geometry* and therefore may be said to quantify the *"topological charge"* of the system, i.e. the helicity of the photon, which is a relativistic quantum number.

Referring to Fig. 3.2.1, the fiber length is

$$s = [p^2 + (2\pi r)^2]^{1/2}, \qquad (3.2.5)$$

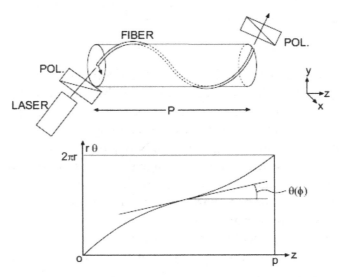

Fig. 3.2.1. (a) Experimental setup; (b) geometry used to calculate the solid angle in momentum space of a nonuniformaly wound fiber on a cylinder. (After Ref. 109.)

and the solid angle in momentum space $\Omega(C)$ spanned by the fiber's closed path C, a circle in the case considered, is

$$\Omega(C) - 2\pi(1 - \cos\theta). \tag{3.2.6}$$

The topological phase is

$$\gamma(C) = -2\pi\sigma\left(1 - \frac{p}{s}\right), \tag{3.2.7}$$

where $\sigma = \pm 1$ is the helicity quantum number of the photon.

By wrapping a piece of paper with a computer-generated curve on a cylinder to which the fiber is fitted, and then unwrapping the paper, the local pitch angle, or the tangent to the curve followed by the fiber, can be estimated to be [Fig. 3.2.1(b)]

$$\theta(\phi) = \tan^{-1}\left(\frac{rd\phi}{dz}\right), \tag{3.2.8}$$

which is the angle between the local wave guide and the helix axes. In momentum space, $\theta(\phi + \pi/2)$ traces out a closed curve C, the fiber path on the surface of a sphere. The solid angle subtended by C to the center of the sphere is

$$\Omega(C) = \int_0^{2\pi} [1 - \cos\theta(\phi)]d\phi. \tag{3.2.9}$$

The topological phase is then, more correctly,

$$\gamma(C) = -\sigma\,\Omega(C) \quad \text{or}$$

$$\gamma(C) = \nu\,\Omega(C) \tag{3.2.10}$$

(where $\nu = 1/2$ in the case of polarization charges, i.e. Pancharatnam's phase). There is thus a linear relation between the angle of rotation of linearly polarized light, and the solid angle $\Omega(C)$ subtended by C at the origin of the momentum space of the photon.[109]

More recently, Chiao *et al.*[129] have demonstrated a topological phase shift in a Mach–Zehnder interferometer in which light travels along nonplanar paths in two arms. They interpret their results in terms of the Aharonov–Anandan phase and changes in projective Hilbert space, i.e. the sphere of spin directions of the photon, rather than parameter (momentum) space. The hypothesis tested was that the evolution of the state of a system is cyclic, i.e. that it returns to its starting point adiabatically or not. Thus the C in Eq. (3.2.10) is to be interpreted as a closed circuit on the sphere of spin directions.

Chiao and Wu[106] consider topological phase rotation effects to be "topological features of the Maxwell theory which originate at the quantum level, but survive the correspondence principle limit $(h \rightarrow 0)$ into the classical level." However, this opinion is contested and the effect is viewed as classical by other authors (e.g. Refs. 102, 103, 130 and 131). For example, the evolution of the polarization vector can be viewed as being determined by a connection on the tangent bundle of the two-dimensional sphere.[130,131] The effect is then viewed as non-Abelian. The situation can then be described with a family of Hamiltonian operators, $H_0 + \mathbf{k} \cdot \mathbf{V}$, where H_0 is rotationally invariant, $\mathbf{V}$ is a vector operator and $\mathbf{k}$ varies over the unit vectors in $\mathbf{R}^3$.[131]

The Chiao–Wu phase and the Pancharatnam phase are additive.[117] This is because the two topological phase effects arise in different parameter spaces: the former in $\mathbf{k}$ space and the latter in polarization vector (Poincaré sphere) space. To see this, Maxwell's equations can be recast into a six-component spinor form[117]

$$\nabla \times (\mathbf{E} \pm \mathbf{B}) = \frac{\pm \partial (\mathbf{E} \pm i\mathbf{B})}{\partial t}, \qquad (3.2.11)$$

$$\Psi = \mathrm{col}(\mathbf{E} + \mathbf{B}, \mathbf{E} - \mathbf{B}) = \mathrm{col}((\Psi^+), (\Psi^-)), \qquad (3.2.12)$$

where **col** denotes a column vector, to obtain a Schrödinger-like equation

$$\frac{i\partial \Psi}{\partial t} = H\Psi, \qquad (3.2.13)$$

where the Hamiltonian, H, is given by

$$H = \begin{matrix} (\nabla x) & 0 \\ 0 & (-\nabla x) \end{matrix} \qquad (3.2.14)$$

(∇x represents the curl). This spinor representation of Maxwell's equations has a natural correspondence with natural optical activity in the frequency domain.[132]

The conventional dynamical phase becomes

$$\delta(H) = -\int_0^T (\Psi, H\Psi)dt \qquad (3.2.15)$$

and the geometrical (topological) phase is, as before

$$\gamma(C) = \oint \mathbf{A} \cdot dl, \qquad (3.2.16)$$

but where the vector potential, $\mathbf{A}$, is explicitly defined as

$$A = -(\Psi, \nabla\Psi), \qquad (3.2.17)$$

i.e. a connection defined on the state space.

By means of Stokes' theorem

$$\gamma(C) = \oint \mathbf{A} \cdot dl = \int_S \nabla \times A dS, \qquad (3.2.18)$$

$$\gamma(C) = \nu \Omega(C), \qquad (3.2.19)$$

as before.

Figure 3.2.2 shows the different manifestations and representations of the topological phase effect. Figure (i) is a sphere of *spin directions* for representing the (Chiao–Wu) phase arising from the spin vector of a photon tracing out a closed curve on the sphere.[117] The topological phase is equal to the angle Ω. Figure (ii) is a Poincaré sphere of polarization states, or *helicity*, of a photon for representing the (Pancharatnam) phase effect arising from cycling in the polarization states of the photon while keeping the direction of the beam fixed.[114] The topological phase is equal to the *negative* of one half the angle Ω. Figure (iii) is a *generalized* Poincaré sphere for representing the angular momentum of light (with space-fixed axis),[117] or a *null flag or twistor* representation[34,133] for representing the Pancharatnam topological phase but with the phase equal to the *positive* value of one half the angle Ω. The topological phase effects represented in (i) and (iii) are additive.

A more explicit relation between the topological phase effect and Maxwell's theory is obtained within the formulation of Maxwell's

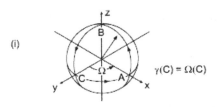

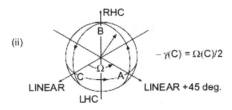

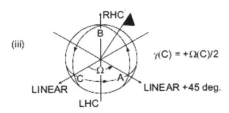

Fig. 3.2.2. (i) A sphere of *spin directions* for representing the (Chiao–Wu) phase arising from the spin vector of a photon tracing out a closed curve on the sphere. (After Ref. 117.) The topological phase is equal to the angle Ω. (ii) A Poincaré sphere of polarization states, or *helicity*, of a photon for representing the (Pancharatnam) phase effect arising from cycling in the polarization states of the photon while keeping the direction of the beam fixed (After Ref. 114). The topological phase is equal to the *negative* of one half the angle Ω. (iii) A *generalized* Poincaré sphere for representing the angular momentum of light (with space-fixed axis) (after Ref. 117), or a *null flag or twistor* representation (after Refs. 34 and 133) for representing the Pancharatnam topological phase but with the phase equal to the *positive* value of one half the angle Ω. The topological phase effects represented in (i) and (iii) are additive.

theory by Biakynicki-Birula and Bialynicka-Birula.[134] Within this formulation, the intrinsic properties of an electromagnetic wave are its wave vector, $\mathbf{k}$, and its polarization, $\mathbf{e}(\mathbf{k})$. As Maxwell's theory can be formulated as a representation of Poincaré symmetry, all wave vectors form a vector space. Thus, implicit in Maxwell's theory is the topology of the surface of a sphere (i.e. the submanifold, S^2). The generators of the Poincaré group involve a covariant derivative in momentum space, whose curvature is given by a magnetic monopole field. However, the Maxwell equations can only determine the polarization tensor, $e(\mathbf{k})$, *up to an arbitrary phase factor*, as Maxwell's theory corresponds to the structure group U(1), i.e. Maxwell's theory *cannot* determine the phase of polarization in momentum space. This arbitrariness permits additional topological phase effects.

On the other hand, the topological phase is *precisely* obtained from a set of angles associated with a group element and there is just one such angle corresponding to a holonomy transformation of a vector bundle around a closed curve on a sphere.[135] The parameter space is the based manifold and each fiber is isomorphic to an N-dimensional Hilbert space. In particular, for the SU(2) case there is a single angle from the holonomy of the Riemannian connection on a sphere.[135] The observation that gauge structure appears in simple dynamical systems — both quantum-mechanical and classical — has been made.[28] For the special case of Fermi systems, the differential geometric background for the occurrence of SU(2) topological phases is the quaternionic projective space with a time evolution corresponding to the SU(2) Yang–Mills instanton.[136] *Locally*, the non-Abelian phase generated can be reduced to an Abelian form. However, it is not possible to define the connection defined on the bundle space except *globally*. This reflects the truly non-Abelian nature of the topological phase. The topological phase effect can be described in a *generalized* Bloch sphere model and an SU(2) Lie group formulation in the spin-coherent state.[137] Furthermore, while acknowledging that in general terms, and formally, Berry's phase is a geometrical object in projective Hilbert space (ray space), the nonadiabatic Berry's phase, physically, is related to the expectation value of spin (spin alignment), and Berry's phase quantization is related to spin alignment quantization.[138]

The topological phase effect even appears in quantum systems constrained by molecular geometry. For example, the topological phase effect appears in the molecular system Na_3.[139] Suppose that a system in an eigenstate $C(r, t)$ responds to slowly varying changes in its parameters $R(t)$, such that the system remains in the *same eigenstate apart from an acquired phase*. If the parameters, $R(t)$, completed a circuit in parameter space, then that acquired phase is not simply the familiar dynamical phase, $[(i\hbar)^{-1}E(R(t)]\,dt$, but an additional geometrical phase factor, $\gamma_n(c)$. The origins of this additional phase factor depend only on the geometry of the parameter space and the topology of the circuit traversed. Therefore, adiabatic excursions of molecular wave functions in the neighborhood of an electronic degeneracy result in a change of phase. That is, if the internuclear coordinates of a wave function traverse a circuit in which the state is degenerate with another, then the electronic wave function acquires an additional phase, i.e. it changes its sign. This change was predicted[140-142] and is a special case of the topological phase applying to a large class of molecular systems exhibiting conical intersections. Delacrétaz *et al.*[139] reported the evidence for half-odd quantization of free molecular pseudorotation and offered the first experimental confirmation of the sign change theorem and a direct measurement of the phase. The topological phase has also been observed in fast-rotating superfluid nuclei, i.e. oscillations of pair transfer matrix elements as a function of the angular velocity[143] and in neutron spin rotation.[144]

The appearance of the topological phase effect in both classical and quantum-mechanical systems thus gives credence to the view that the $\mathbf{A}_\mu$ potentials register physical effects at both the classical and the quantum-mechanical level. That such a role for these potentials exists at the quantum-mechanical level is not new. *It is new to consider the $\mathbf{A}_\mu$ potentials for such a role at the classical level.* One may ask how the schism in viewing the $\mathbf{A}_\mu$ potentials came about, i.e. why are they viewed as physical constructs in quantum mechanics, but as merely arbitrary mathematical conveniences in classical mechanics? *The answer is that whereas quantum theory is defined with respect to boundary conditions, in the formal presentation of Maxwell's theory boundary conditions are undefined.* Stokes' theorem demonstrates this.

3.3. *Stokes' theorem re-examined*

Stokes' theorem of potential theory applied to classical electromagnetism relates diverging potentials on line elements to rotating potentials on surface elements. Thus, Stokes' theorem describes a local-to-global field relationship.

If $\mathbf{A}(x)$ is a vector field, S is an open, orientable surface, C is the closed curve bounding S, $\mathbf{dl}$ is a line element of C, $\mathbf{n}$ is the normal to S and C is traversed in a right-hand screw sense (positive direction) relative to $\mathbf{n}$, then the line integral of A is equal to the surface integral over S of $(\nabla \times \mathbf{A}) \cdot \mathbf{n}$:

$$\oint \mathbf{A} \cdot \mathbf{dl} = \int_S (\nabla \times \mathbf{A}) \cdot \mathbf{n}\, da. \qquad (3.3.1)$$

It is also necessary that S be the union of a finite number of smooth surface elements and that the first order partial derivation of the components of $\mathbf{A}$ be continuous on S. Thus Stokes' theorem, as described, takes no account of: (i) space–time overlap in a region with fields derived from different sources; (ii) the exact form of the boundary conditions.

This neglect of the exact form of the boundary conditions in Stokes' theorem of classical mechanics can be contrasted with the situation in quantum mechanics. In quantum mechanics, the wave function satisfies a partial differential equation coupled to boundary conditions because the Schrödinger equation describes a minimum path solution to a trajectory between two points. The boundary condition in the doubly connected (overlap) region outside of the shielded volume in an AB experiment is the reason for the single-valuedness of the wave function, and also the reason for quantization. The situation is also different with spatial symmetries other than the usual, Abelian, spatial symmetry.

A non-Abelian Stokes theorem is[145]

$$h^{-1}\left(\frac{dh}{ds}\right) \approx ie \int_0^1 g^{-1}\mathbf{G}_{ij}g\left(\frac{\partial r^i}{\partial t}\right)\left(\frac{\partial r^j}{\partial s}\right) dt, \qquad (3.3.2)$$

where $h(s)$ is a path-dependent phase factor associated with a closed loop and defines a closed loop $r(s, t)$, $0 \le t \le 1$, s fixed, in the U(1) symmetry space, $\mathbf{H}$ (equivalent to $\mathbf{A}_\mu$); G is a gauge field tensor for the SU(2) non-Abelian group; and g is magnetic charge. Here, the boundary conditions, i.e. the path dependencies, are made explicit, and we have a local field [with U(1) symmetry] to global field [with SU(2) symmetry] connection.

In classical electromagnetism, therefore, Stokes' theorem appears merely as a useful mathematical relation between a vector field and its curl. *In gauge theory, on the other hand, an amended Stokes' theorem would provide the value for the net comparative phase change in the internal direction of a particle traversing a closed path, i.e. a local-to-global connection.*

Lest it be thought that the A_μ field which functions as the independent variable in the AB experiment is only a quantum effect with no relevance to classical behavior, the relation of the A_μ potential to the properties of bulk condensed matter is examined in the following section. A more complete definition of Stokes' theorem is also given in Sec. 4 below [Eq. (4.10)].

Use of Stokes' theorem has a price: that of the (covert) adoption of a gauge for local-to-global connections. This is because Stokes' theorem applies directly to propagation issues, which are defined by local-to-global connections. Such connections are also required in propagation through matter. Thus, there is a requirement for Stokes' theorem in any realistic definition of macroscopic properties of matter, and in the next section we see that the physical effects of the A_μ potentials exist not merely in fields traversing through various connecting topologies, but in radiation–matter interactions.

3.4. *Properties of bulk condensed matter — Ehrenberg and Siday's observation*

In the AB F $\rightarrow$ FE situation, when the size of the solenoid is much larger than the de Broglie wavelength of the incident electrons, the scattering amplitude is essentially dominated by simple classical trajectories. But the classical manifestation of quantum influences is not peculiar to the AB effect. For example, macroscopic quantum tunneling is observable in Josephson tunnel junctions in which the phase difference of the junction can be regarded as a macroscopic degree of freedom, i.e. a classical variable.[146,147]

Even without known quantum influences or quantum-mechanical explanation, there is a classical justification for the A_μ potential as a physical effect. For example, on the basis of optical arguments, the A_μ potential must be chosen so as to satisfy Stokes' theorem, *thereby removing the arbitrariness with respect to the gauge.* Furthermore, an argument originating with Ehrenberg and Siday[148] shows that a gauge-invariant A_μ potential is *presupposed* in any definition of the refractive index.

This argument is a derivation of the refractive index based on *Fermat's principle*: in any optical medium, a scalar quantity, e.g. the refractive index, finite everywhere in space, can be defined so that the line integral in the three-dimensional space taken between any fixed points must be an extremum which passes through these points. The optical path along a given line connecting a point 1 and a point 2 is

$$\int_1^2 m\,ds = \int_1^2 [mv + (A_\mu \cdot n)]\,ds, \qquad (3.4.1)$$

where n is the unit vector in the direction of the line, v is the velocity of the electron, and m is its mass. Defined in this way, an unambiguous definition of the refractive index indicates the *necessity* of a unique (gauge-invariant) definition of the A_μ potential. Stated differently: an unambiguous definition of the refractive index implies defining the boundary conditions through which test radiation moves. These boundary conditions define a definite gauge and thereby definite A_μ potentials.

This an example of physical A_μ -dependent effects (the refractive properties of matter) seen when radiation propagates through matter — from one point to another. In the next section A_μ effects are described when two fields are in close proximity. This is the Josephson effect, and again, the potential functions as a global-to-local operator.

3.5. *The Josephson effect*

Josephson[149–152] predicted that a d.c. voltage, V, across the partitioning barrier of a superconductor gives rise to an alternating current of frequency

$$\omega = \frac{2eV}{\hbar}. \qquad (3.5.1)$$

The equivalent induced voltage is[153]

$$V = \left(\frac{1}{c}\right)\frac{d\Phi}{dt}, \qquad (3.5.2)$$

where Φ is the magnetic flux through a superconducting ring containing a barrier. The circulating current, I, exhibits a periodic dependence upon Φ

$$I(\alpha) = \sum_n a_n \sin 2\pi\alpha, \qquad (3.5.3)$$

where

$$\alpha = \frac{\Phi}{\frac{\hbar c}{e}}. \qquad (3.5.4)$$

The validity of Eq. (3.5.4) depends upon the substitution of

$$\mathbf{p} - \frac{e\mathbf{A}}{c} \qquad (3.5.5)$$

for the momentum, $\mathbf{p}$, of any particle with charge and with a required gauge invariance for the $\mathbf{A}$ potential.

The phase factor existing in the junction gap of a Josephson junction is an exponential of the integral of the $\mathbf{A}$ potential. The fluxon, or the decrementlessly conducting wave in the long Josephson junction and in a SQUID, is the equivalent of an $\mathbf{A}$ wave in one-dimensional phase space. The phenomenological equations are

$$\frac{\partial\phi}{\partial x} = \left(\frac{2ed}{\hbar c}\right)H_y, \qquad (3.5.6)$$

$$\frac{\partial\phi}{\partial t} = \left(\frac{2e}{\hbar}\right)V, \qquad (3.5.7)$$

$$J_x = j\sin\phi + \sigma V, \qquad (3.5.8)$$

where ϕ is the phase difference between two superconductors; H is the magnetic field in the barrier; V is the voltage across the barrier; $d = 2\lambda + l$; λ is the penetration depth; and l is the barrier thickness.[g]

If the barrier is regarded as having a capacitance, C, per unit area, then Eq. (3.5.6) and Maxwell's equations give

$$\left[\frac{\partial^2}{\partial x^2} - \left(\frac{1}{c^2} \right) \left(\frac{\partial^2}{\partial t^2} \right) - \left(\frac{\beta}{c^2} \right) \left(\frac{\partial}{\partial t} \right) \right] \phi = \left(\frac{1}{\lambda_0^2} \right) \sin \phi, \qquad (3.5.9)$$

where $c^2 = c^2/4\pi dC$ is the phase velocity in the barrier, $\lambda_0^2 = hc^2/8\pi edj$ is the penetration depth and $\beta = 4\pi dc^2\sigma = \sigma/C$ is the damping constant. Anderson[250] demonstrated that solutions to this equation, representing vortex lines in the barrier, are obtained as solutions to

$$\frac{\partial \phi^2}{\partial x^2} = \left(\frac{1}{\lambda_0^2} \right) \sin \phi, \qquad (3.5.10)$$

which, except for the sign, is the equation of a pendulum.

The Josephson effect is remarkable in the present context for three reasons: (i) with well-defined boundary conditions (the barrier), the phase, ϕ, is a well-defined gauge-invariant variable; (ii) an equation of motion can be defined in terms of the well-studied pendulum,[154] relating a phase variable to potential energy; (iii) the "free" energy in the barrier is[155]

$$F = \left(\frac{\hbar j}{2e} \right) \int dx \left[(1 - \cos \phi) + \frac{1}{2\lambda_0^2} \left(\frac{\partial \phi}{\partial x} \right)^2 + \frac{1}{2} \left(\frac{\lambda_0}{c} \right)^2 \left(\frac{\partial \phi}{\partial t} \right)^2 \right],$$

$$(3.5.11)$$

an equation which provides an (free) energy measure in terms of the differential of a phase variable. The Josephson effect, like the AB effect, demonstrates the registration of physical influences by means of phase changes. The Josephson phase, also like the AB phase, registers field influences.

---

[g]Lenstra *et al.*[254] have shown an analogy between Josephson-like oscillations and the Sagnac effect.

Jaklevic *et al.*[156] studied multiply connected superconductors utilizing Josephson junction tunneling and modulated the supercurrent with an applied magnetic field. The interference "fringes" obtained were found to occur even when the magnetic flux is confined to a region not accessible to the superconductor, i.e. there occurs vector potential modulation of superconducting electron drift velocity. As always, the superconductive state had global phase coherence, indicating that the modulation effect studied was a local $(\mathbf{A}_\mu)$ influence on global phase effects (i.e. the phase order parameter in the barrier).

In the case of the next effect examined, the quantized Hall effect, the effect is crucially dependent upon the gauge invariance of the $\mathbf{A}_\mu$ potential. The result of such gauge invariance is remarkably significant: *an independence of the quantization condition on the density of mobile electrons in a test sample.*

This independence was seen above while examining the remarkable independence in preservation of phase coherence in electrons over distances larger than the atomic spacing or the free path length in the $F \rightarrow CE$ AB effect. In both cases, the primacy of macroscopic and "mesoscopic," effects are indicated.

3.6. *The quantized Hall effect*

The quantized Hall effect[157,158] has the following attributes:

(1) There is the presence of a Hall conductance σ_{xx} in a two- dimensional gas within a narrow potential well at a semiconductor — heterostructure interface e.g. in MOS, quantum well and MOSFET;

(2) The temperature is low enough that the electrons are all in the ground state of the potential well and with the Fermi level being between the Landau levels;

(3) The conductance is quantized with a plateau having $\sigma_{xy} = n\hbar/e^2\sigma_{xy}$ (n is an integer) for finite ranges of the gate voltage in which the regular conductance is severely reduced;

(4) Together with the well-known Hall effect (1879) condition (a magnetic field perpendicular to the plane and an electric field in the plane and the electrons drifting in the direction $\mathbf{E} \times \mathbf{B}$), the

energy associated with the cyclotron motion of each electron takes on quantized values $(n + 1/2)\hbar\omega_c$, where ω_c is the cyclotron frequency at the imposed magnetic field and n is the quantum number corresponding to the Landau level.

The AB flux, or $\mathbf{A}$ wave, can be generated in such two-dimensional systems and be increased by one flux quantum by changing the phase of the ground state wave function around the system. The quantized Hall effect is thus a macroscopic quantum Hall phenomenon related to the fundamental role of the phase and the $\mathbf{A}_\mu$ potential in quantum mechanics.

An important feature of the quantized Hall effect is the lack of dependence of quantization (integral multiples of $e^2/\hbar$) on the density of the mobile electrons in the sample tested (but, rather, on the symmetry of the charge density wave[159]). Underlying this lack of dependence is a required gauge invariance of the $\mathbf{A}_\mu$ potential. For example, the current around a metallic loop is equal to the derivative of the total electronic energy, U, of the system with respect to the magnetic flux through the loop, i.e. with respect to the $\mathbf{A}_\mu$ potential pointing around the loop[160]

$$I = \frac{(c/L)\partial U}{\partial \mathbf{A}}. \tag{3.6.1}$$

As this derivative is nonzero only with phase coherence around the loop, i.e. with an extended state, Eq. (3.6.1) is valid only if

$$A = \frac{n\hbar c}{eL}, \tag{3.6.2}$$

i.e. only with a gauge invariance for $\mathbf{A}$.

With a gauge invariance defined for $\mathbf{A}$, and with the Fermi level in a mobility gap, a vector potential increment changes the total energy, U, by forcing the filled states toward one edge of the total density of states spectrum and the wave functions are affected by a vector potential increment only through the location of their centers. Therefore, *gauge invariance of the A potential, being an exact symmetry, forces the addition of a flux quantum to result in only an excitation or de-excitation of the total system.*[160] Furthermore, the energy gap exists *globally* between the electrons and holes affected by such a perturbation in the way described, rather than in specific local density of states. Thus, the Fermi level lies *globally* in a gap in an extended state spectrum and there is no dependence of Hall conductivity on the density of mobile electrons.

Post[161–163] has also implied the vector potential in the conversion of a voltage–current ratio of the quantized Hall effect into a ratio of period integrals. If V is the Hall voltage observed transversely from the Hall current I, the relation is

$$\frac{V}{I} = \frac{\oint \mathbf{A}}{\oint \mathbf{G}} = Z_H = \text{quantized Hall impedance,} \tag{3.6.3}$$

where $\mathbf{G}$ defines the displacement field $\mathbf{D}$ and the magnetic field $\mathbf{H}$. The implication is that

$$\frac{V}{I} = \frac{\int_0^T V dt}{\int_0^T I dt}, \tag{3.6.4}$$

where

$$\int_0^T V dt = \oint \mathbf{A}, \quad \text{the quantization of magnetic flux,} \tag{3.6.5}$$

$$\int_0^T I dt = \oint \mathbf{G}, \quad \text{the quantization of electric flux,} \tag{3.6.6}$$

and T is the cyclotron period.

Aoki and Ando[164] also attribute the universal nature of the quantum Hall effect, i.e. the quantization in units of $e^2/\hbar$ at $T = 0$ for every energy level in a finite system, to a topological invariant in a mapping from the gauge field to the complex wave function. These authors assume that in the presence of external AB magnetic fluxes, the vector potential $\mathbf{A}_0$ is replaced by $\mathbf{A}_0 + \mathbf{A}$, where $\mathbf{A} = (A_x, A_y)$. In cylindrical geometry, a magnetic flux penetrates the opening of the cylinder and the vector potential is thought of as two magnetic fluxes, $(\mathbf{\Phi}_x, \mathbf{\Phi}_y) = (A_x L, A_y L)$, penetrating inside and through the opening of a torus when periodic boundary conditions are imposed in both the x and y directions for a system of size L. According to the Byers–Yang theorem,[165] the physical system assumes its original state when A_x or A_y increases by Φ_0/L, where $\Phi_0 = \hbar c/e$, the magnetic flux quantum.

The next effect examined, the de Haas–van Alphen effect, also pivots on $\mathbf{A}_\mu$ potential gauge invariance.

3.7. *The de Haas–van Alphen effect*

In 1930, de Haas and van Alphen observed what turned out to be susceptibility oscillations with a changing magnetic field which were periodic with the reciprocal field. Landau showed in the same year that for a system of free electrons in a magnetic field, the motion of the electrons *parallel* to the field is *classical*, while the motion of the electrons *perpendicular* to the field is *quantized*; and Peierls showed in 1933 that this holds for free electrons in a metal (with a spherical Fermi surface). Therefore, the free energy of the system and thus the magnetic moment ($M = \partial F/\partial H$) oscillates with the magnetic field H. This oscillation is the major cause of the De Haas–Van Alphen effect.

In 1952, Onsager showed that the frequencies of oscillations are directly proportional to the extremal cross-sections of the Fermi surface perpendicular to the magnetic field. If $\mathbf{p}$ is the electronic momentum and

$$\left[p - \left(\frac{e}{c} \right) \mathbf{A} \right] \tag{3.7.1}$$

is the canonical momentum [cf. Eqs. (3.5.5)], then

$$\oint \left[p - \left(\frac{e\mathbf{A}}{c} \right) \cdot dl \right] = (n + \gamma)h, \tag{3.7.2}$$

where n is an integer and γ is a phase factor. The relation of the $\mathbf{A}$ vector potential and the real space orbit is

$$\oint \mathbf{A} \cdot dl = \int \nabla \times \mathbf{A} \cdot dS = \mathbf{H}S, \tag{3.7.3}$$

where S is the area of the orbit in real space. Furthermore, electron paths in momentum space have the same shape as those in real space but changed in scale and turned through 90°, due to the Lorentz force relation: $d\mathbf{p}/dt = (e/c)(\nabla \times \mathbf{H})$.

Therefore, as (i) the area of the orbit in momentum space is $S = (n + \gamma)(e\hbar H/c)$ and (ii) the susceptibility is $-(1/H)(\partial F/\partial H)$, which is periodic in $1/H$ with period $\Delta(1/H) = 2\pi e/c\hbar S$, there is a direct influence of the **A** vector potential on the de Haas–van Alphen effect due to the phase factor dependence [Eq. (3.7.2)]. Thus the validation of Eqs. (3.7.1) and (3.7.2) requires A_μ potential gauge invariance. (The relation between the AB effect and the quantized Hall effect has been observed by Timp *et al.*[71])

Two effects have now been examined pivoting on A_μ potential gauge invariance. *This gauge invariance implies flux conservation, i.e. a global conservation law.* The next effect examined, the Sagnac effect, makes explicit the consequences of this global conservation.

3.8.   *The Sagnac effect*

In 1913, Sagnac demonstrated a fringe shift by rotating an interferometer (with a polygonal interference loop traversed in opposite senses) at high speed[166–168] (Fig. 3.8.1). Einstein's general theory of relativity predicts a phase shift proportional to the angular velocity and to the area enclosed by the light path — not because the velocity of the two

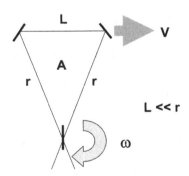

Fig. 3.8.1.   The Sagnac interferometer in which the center of rotation coincides with the beam splitter location. The Sagnac phase shift is independent of the location of the center of rotation and the shape of the area. The phase shift along L is independent of r. (After Silvertooth.[169])

beams is different, but because they each have their own time. However, *the AB, AAS and topological phase effects deny Lorentz invariance to the electromagnetic field as any field's natural and inevitable implication, i.e. Lorentz invariance is not "built in" to the Maxwell theory — it is a gauge implied by special A_μ potential conditions, i.e. special boundary conditions imposed on the electromagnetic field.* Therefore, the Einstein interpretation pivots on unproven boundary conditions and the effect is open to other, competing, explanations (cf. Ref. 170).

A different explanation is offered by the Michelson *et al.*[171–173] experiments of 1924–1925. These investigators predicted a phase shift more simply on the basis of a difference in the velocity of the counter propagating beams and the earth rotating in a stationary ether without entrainment. (It should be noted that the beam path in the well-known Michelson–Morley 1886 interferometer[174] does not enclose a finite surface area. Therefore this experiment cannot be compared with the experiments and effects examined in the present review, and in fact, according to these more recent experiments, no fringe shift can be expected as an outcome of a Michelson–Morley experiment, i.e, the experiment was not a test for the presence of an ether.[h])

Post[175] argues that the Sagnac effect demonstrates that the space–time formulations of the Maxwell equations do not make explicit the constitutive properties of free space. The identification $\mathbf{E} = \mathbf{D}, \mathbf{H} = \mathbf{B}$, in the absence of material polarization mechanisms in free space, is the so called *Gaussian field identification.*[176] This identification is equivalent to an unjustified adoption of Lorentz invariance. However, the Sagnac effect and the well-used ring laser gyro on which it is based indicate that in a rotating frame, the Gaussian identity does not apply. This requirement of metric independence was proposed by Van Dantzig.[177] In order to define the constitutive relations between the fields $\mathbf{E}$ and $\mathbf{B}$

---

[h]The author is indebted to an anonymous reviewer for indicating the paper by Post[181] which shows that not only does the Michelson and Morley experiment not disprove the existence of an ether, but that the experiment should give a null result regardless of whether the motion is uniform or not. Post also demonstrates a mutual relation between the Michelson–Morley and Sagnac experiments.

constituting a covariant six-vector $\mathbf{F}_{\lambda\nu}$, and the fields $\mathbf{D}$ and $\mathbf{H}$, constituting a contravariant six-vector, $\mathbf{G}^{\lambda\nu}$, the algebraic relation[176]

$$\mathbf{G}^{\lambda\nu} = \frac{1}{2}\chi^{\lambda\nu\sigma\kappa}\mathbf{F}_{\lambda\nu} \tag{3.8.1}$$

was proposed, where $\chi^{\lambda\nu\sigma\kappa}$ is the constitutive tensor and Eq. (3.8.1) is the constitutive map. The generally invariant vector d'Alembertian (wave equation) is

$$\partial_\nu\chi^{\lambda\nu\sigma\kappa}\partial_\sigma\mathbf{A}_\kappa = 0, \tag{3.8.2}$$

indicating the vector potential dependence.

The pivotal role of the vector potential is due to the flux conservation, which is a *global* conservation law.[161,178] The *local* conservation law of flux

$$d\mathbf{F} = 0, \tag{3.8.3}$$

excludes a role for the $\mathbf{A}$ potential ($\mathbf{F}$ is inexact). However, only if

$$\oint \mathbf{F} = 0 \tag{3.8.4}$$

is it possible to state that $d\mathbf{A} = \mathbf{F}$ ($\mathbf{F}$ is exact). In other words, $d\mathbf{F} = 0$ implies $\oint \mathbf{F} = 0$ *only if the manifold over which $\mathbf{F}$ is defined is compact and simply connected*, e.g. one-connectedness (contractable circles), two-connectedness (contractable spheres) and three-connectedness (contractable three-spheres).

Post[179–181] argues that the constitutive relations of the medium-free fields $\mathbf{E}$ and $\mathbf{H}$ to the medium left out treatment of free space as a "medium." If $\mathbf{C}$ is the differential three-form of charge and current density, then the *local* conservation of charge is expressed by

$$d\mathbf{C} = 0, \tag{3.8.5}$$

and the *global* definition is

$$\mathbf{C} = d\mathbf{G}. \tag{3.8.6}$$

The Post relation is in accord with the symmetry of space–time and momentum–energy required by the reciprocity theory of Born[182] and, more recently, that of Ali.[183,184] Placing these issues in a larger context, Hayden[185] has argued that the classical interpretation of the Sagnac effect indicates that the speed of light is not constant. Furthermore, Hayden,[186] emphasizing the distinction between kinematics (which concerns space and time) and mechanics (which concerns mass, energy, force and momentum), argues that time dilation (a kinematic issue) is difficult to distinguish from changes in mass (a mechanics issue) owing to the way time is measured. The demonstration of an analogy between Josephson-like oscillations and the Sagnac effect[254] supports this viewpoint.

3.9. *Summary*

In summary, the following effects have been examined:

(i) The Aharonov–Bohm and Altshuler–Aronov–Spival effects, in which changes in the A_μ potential at a third location indicate differences in the A_μ field along *two* trajectories at *two* other locations.

(ii) The topological phase effects, in which changes in the spin direction or polarization defined by the A_μ potential at one location, *a*, are different from that at another location, *b*, owing to topological winding of the trajectory between the *two* locations *a* and *b*.

(iii) Stokes' theorem, which requires precise boundary conditions for *two* fields — the local and global fields — for exact definition in terms of the A_μ potential.

(iv) Ehrenberg and Siday's derivation of the refractive index, which describes propagation between *two* points in a medium and which requires gauge invariance of the A_μ potential.

(v) The Josephson effect, which implies the A_μ potential as a local-to-global operator connecting *two* fields.

(vi) The quantized Hall effect, which requires gauge invariance of the A_μ potential in the presence of *two* fields.

(vii) The de Haas–van Alphen effect, which requires gauge invariance of the A_μ potential in the presence of *two* fields.

(viii) The Sagnac effect, which requires flux conservation, i.e. gauge invariance of the $\mathbf{A}_\mu$ potential in comparing *two* fields before and after movement.

All these effects pivot on a physical definition of $\mathbf{A}_\mu$ potentials. In the next section, the theoretical reasons for questioning the completeness of Maxwell's theory are examined as well as the reasons for the physical effectiveness of the $\mathbf{A}_\mu$ potentials in the presence of two fields. The $\mathbf{A}_\mu$ potentials have an ontology or physical meaning as *local* operators mapping *local* e.m. fields onto *global* spatiotemporal conditions. This operation is measurable if there is a second comparative mapping of the local conditioned fields in a many-to-one fashion (multiple connection).

4.   Theoretical Reasons for Questioning the Completeness of Maxwell's Theory

Yang[27,98,187,188] interpreted the electromagnetic field in terms of a nonintegrable (i.e. path-dependent) phase factor by an examination of Dirac's monopole field.[189,190] According to this interpretation, the AB effect is due to the existence of this phase factor whose origin is due to the topology of connections on a fiber bundle.

The *phase factor*

$$\exp\left[\frac{ie}{\hbar c}\oint \mathbf{A}_\mu dx^\mu\right],\tag{4.1}$$

according to this view, is physically meaningful, but not the *phase*

$$\left(\frac{ie}{\hbar c}\oint \mathbf{A}_\mu dx^\mu\right),\tag{4.2}$$

which is ambiguous because different phases in a region may describe the same physical situation. The phase factor, on the other hand, can distinguish different physical situations having the same field strength but different action. Referring to Fig. 4.1, the phase factor for any

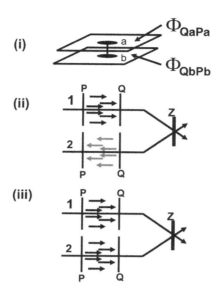

Fig. 4.1.   (i) The overlap area [Z in (ii) and (iii)] showing a mapping from location a to b. The phase factor Φ_{QaPa} is associated with the e.m. field which arrived at Z through path 1 in (ii) and (iii) and Φ_{QbPb} with the e.m. field which arrived at Z through path 2 in (ii) and (iii). (ii) In paths 1 and 2 the e.m. fields are conditioned by an **A** field between P and Q oriented in the direction indicated by the arrows. Note the reversal in direction of the **A** field in paths 1 and 2, hence $S(P) \neq S(Q)$ and $\Phi_{QaPa} \neq \Phi_{QbPb}$. (iii) Here the conditioning **A** fields are oriented in the same direction, hence there is no noticeable gauge transformation and no difference noticeable in the phase factors $S(P) = S(Q)$ and $\Phi_{QaPa} = \Phi_{QbPb}$. (After Ref. 98.)

path from, say, P to Q, is

$$\Phi_{QP} = \exp\left[\frac{ie}{\hbar c} \int_Q^P \mathbf{A}_\mu dx^\mu\right]. \tag{4.3}$$

For a *static magnetic monopole* at an origin defined by the spherical coordinate, $r = 0, \theta$ with azimuthal angle ϕ, and considering the region R of all space–time *other than this origin*, the gauge transformation in the overlap of two regions, a and b, is

$$S_{ab} = \exp(-i\alpha) = \exp\left[\left(\frac{2ige}{\hbar c}\right)\phi\right], \quad 0 \leq \phi \leq 2\pi, \tag{4.4}$$

where g is the monopole strength.

This is an allowed gauge transformation if and only if

$$\frac{2ge}{\hbar c} = \text{an integer} = D, \tag{4.5}$$

which is Dirac's quantization. Therefore

$$S_{ab} = \exp(iD\phi). \tag{4.6}$$

In the overlapping region, there are two possible phase factors, Φ_{QaPa} and Φ_{QbPb}, and

$$\Phi_{QaPa} S(P) = S(Q)\Phi_{QbPb}, \tag{4.7}$$

a relation which states that $(A_\mu)_a$ and $(A_\mu)_b$ are related by a gauge transformation factor.

The general implication is that for a gauge with any field defined on it, the total magnetic flux through a sphere around the origin $r = 0$ is independent of the gauge field and only depends on the gauge (phase)

$$\int\int \phi_{\mu\nu}dx^\mu dx^\nu = \left(\frac{-i\hbar c}{e}\right)\int \frac{\partial}{\partial x^\mu}(\ln S_{ab})dx^\mu, \tag{4.8}$$

where the integral is taken around any loop around the origin $r = 0$ in the overlap between the R_a and R_b, as, for example, in an equation for a sphere $r = 1$. As S_{ab} is single-valued, this integral must be equal to an integral multiple of a constant (in this case $2\pi i$).

Another implication is that if the A_μ potentials originating from, or passing through, two or more different local positions are gauge-invariant when compared to another, again different, local position, then the referent providing the basis or metric for the comparison of the phase differences at this local position is a unit magnetic monopole. The unit monopole, defined at $r = 0$, is unique in not having any internal degrees of freedom.[191] Furthermore, both the monopole and charges are topologically conserved, but whereas *electric* charge is topologically conserved in U(1) symmetry, *magnetic* charge is only conserved in SU(2) symmetry.

Usually, there is no need to invoke the monopole concept as the A_μ field is, as emphasized here, treated as a mathematical, not physical, construct in contemporary classical physics. However, in quantum physics, the wave function satisfies a partial differential equation coupled to boundary conditions. The boundary conditions in the doubly connected region outside of the solenoid volume in an AB experiment result in the single-valuedness of the wave function, which is the reason for quantization. Usually, for example in textbooks explaining the theory of electromagnetism as noted above, Stokes' theorem is written as

$$\oint \mathbf{A} dx = \int \int \mathbf{H} \cdot ds = \int_S (\nabla \times \mathbf{A}) \cdot \mathbf{n} \, da, \qquad (4.9)$$

and no account is taken of space–time overlap of regions with fields derived from different sources having undergone different spatiotemporal conditioning, and no boundary conditions are taken into account. Therefore, no quantization is required.

There is no lack of competing opinions on what the theoretical basis is for the magnetic monopole implied by gauge-invariant $\mathbf{A}_\mu$ potentials[145,192–195] (cf. Ref. 196 for a guide to the literature). The Dirac magnetic monopole is an anomalously shaped (string) magnetic dipole at a singularity.[189,190] The Schwinger magnetic monopole is essentially a double singularity line.[197,198] However, gauge-invariant $\mathbf{A}_\mu$ potentials are the local manifestation of global constructs. This precludes the existence of *isolated* magnetic monopoles, but permits them to exist *globally* in any situation with the requisite energy conditioning. Wu and Yang,[98,199,200] 't Hooft,[201,202] Polyakov[203] and Prasad and Sommerfeld[204] have described such situations.

More recently, Zeleny[205] has shown that Maxwell's equations and the Lorentz force law can be derived, not by using invariance of the action (Hamilton's principle) or by using constants of the motion (Lagrange's equations), but by considerations of symmetry. If, in the derivation, the classical $\mathbf{A}$ field is dispensed with in favor of the electromagnetic tensor $\mathbf{F}$, classical magnetic monopoles are obtained, which are without strings and can be extended particles. Such particles are accelerated by a magnetic field and bent by an electric field.

Related to mechanisms of monopole generation is the Higgs field (Φ) approach to the vacuum state.[206–208] The field, in some scenarios, breaks a higher order symmetry field, e.g. SU(2), **G**, into **H** of U(1) form. The **H** field is then proportional to the electric charge.

There are at least five types of monopoles presently under consideration:

(1) The Dirac monopole, a point singularity with a string source. The A_μ field is defined everywhere except on a line joining the origin to infinity, which is occupied by an infinitely long solenoid, so that $\mathbf{B} = \nabla \cdot \mathbf{A}$ (a condition for the existence of a magnetic monopole). Dirac's approach assumes that a particle has either electric or magnetic charge but not both.

(2) Schwinger's approach, on the other hand, permits the consideration of particles with both electric and magnetic charge, i.e. dyons.[197,198]

(3) The 't Hooft–Polyakov monopole, which has a smooth internal structure but without the need for an external source. There is, however, the requirement for a Higgs field.[206–208] The 't Hooft–Polyakov model can be put in the Dirac form by a gauge transformation.[145]

(4) The Bogomol'nyi–Prasad–Sommerfeld monopole is one in which the Higgs field is massless, long range and with a force which is always attractive.

(5) The Wu–Yang monopole requires no Higgs field, has no internal structure and is located at the origin. It requires multiply connected fields. The singular string of the Dirac monopole can be moved arbitrarily by a gauge transformation.[209] Therefore, the Dirac and the Wu–Yang monopoles can be made compatible. The Higgs field formalism can also be related to that of Wu–Yang, in which only the exact symmetry group appears.

Goddard and Olive[145] demonstrated that there are two conserved currents for a monopole solution: the usual Noether current, whose conservation depends on the equations of motion; and a topological current, whose conservation is independent of the equations of motion.

Yang[210] showed that if space–time is divided into two overlapping regions in both of which there is a vector potential **A** with gauge transformation between them in the overlap regions, then the proper definition of Stokes' theorem when the path integral goes from region 1 to another 2 is[199,200]

$$\int_A^C \mathbf{A}dx = \int_A^B \mathbf{A}_1 dx = \int_B^C \mathbf{A}_2 dx + \beta(B). \tag{4.10}$$

The β function is defined by the observation that in the region of overlap, the difference of the vector potentials $\mathbf{A}_1 - \mathbf{A}_2$ is curl-less as the two potentials give the same local electromagnetic field.

There are also general implications. Gates[211] takes the position that all the fundamental forces in nature arise as an expression of gauge invariance. If a phase angle $\theta(x, t) = -(i/2)\ln[\psi/\hat{\psi}]$ is defined for quantum-mechanical systems, then although the difference $\theta(x_1, t) - \theta(x_2, t)$ is a *gauge-dependent* quantity, the expression

$$\theta(x_1, t) - \theta(x_2, t) + \left(\frac{e}{\hbar c}\right) \int_{x_1}^{x_2} ds\, \mathbf{A}(s, t) \tag{4.11}$$

is *gauge-invariant*. (Note that according to the Wu–Yang interpretation, the last expression should be $\exp[(e_0/\hbar c) \int_{x_1}^{x_2} ds\, \mathbf{A}(s, t)]$. Therefore, any measurable quantity which is a function of such a difference in phase angles must also depend on the vector potentials shown. Setting the expression (4.11) to zero gives a general description of both the AB and Josephson effects. Substituting $\exp[(e_0/\hbar c) \int_{x_1}^{x_2} ds\, \mathbf{A}(s, t)]$ for the final term gives a description of the topological phase effect.

The phenomena described above are a sampling of a range of effects. There are probably many yet to be discovered, or provided with the description of an "effect." A unifying theme of all of them is that the physical effect of the $\mathbf{A}_\mu$ potentials is only describable (a) when *two* or more fields undergo different spatiotemporal conditioning *and* there is also a possibility of cross-comparison (many-to-one mapping) or, equivalently, (b) in the situation of a field trajectory with a *beginning* (giving the field before the spatiotemporal conditioning) and an *end* (giving the field after the conditioning), *and* again a possibility of cross-comparison. Setting boundary conditions to an e.m. field gives gauge invariance but without necessarily providing the conditions for detection of the gauge invariance. The gauge-invariant $\mathbf{A}_\mu$

potential field operates on an e.m. field state to an extent determined by global symmetries defined by spatiotemporal conditions, but the effect of this operation or conditioning is only detectable under the global conditions (a) and (b). With no interfield mapping or comparison, as in the case of the solitary electromagnetic field, the A_μ fields remain ambiguous, but this situation occurs only if no boundary conditions are defined — an ambiguous situation even for the electric and magnetic fields. Therefore, the A_μ potentials in all useful situations have a meaningful physical existence related to boundary condition choice — even when no situation exists for their comparative detection. *What is different between the A_μ field and the electric and magnetic fields is that the ontology of the A_μ potentials is related to the global spatiotemporal boundary conditions in a way in which the local electric and magnetic fields are not.* Owing to this global spatiotemporal (boundary condition) dependence, the operation of the A_μ potentials is a one-to-many, local-to-global mapping of individual e.m. fields, the nature of which is examined in Subsec. 5.2. The detection of such mappings is only within the context of a *second* comparative projection, but this time global-to-local.

This section has addressed theoretical reasons for questioning the completness of U(1) symmetry, or Abelian Maxwell theory in the presence of *two* local fields separated globally. In the next section, a pragmatic reason is offered: propagating velocities of e.m. fields in lossy media cannot be calculated in U(1) Maxwell theory. The theoretical justification for physically defined A_μ potentials lies in the application of Yang–Mills theory — not to high energy fields, where the theory first found application, but to low energy fields crafted to a specific group of transformation rules by boundary conditions. This is a new application of Yang–Mills theory.

5. Pragmatic Reasons for Questioning the Completeness of Maxwell's Theory

5.1. *Harmuth's ansatz*

A satisfactory concept permitting the prediction of the propagation velocity of e.m. signals does not exist within the framework of Maxwell's theory[212–214] (see also Refs. 215–223). The calculated group velocity fails for two reasons: (i) it is almost always larger than the speed of light for RF transmission through the atmosphere; (ii) its derivation implies a transmission rate of information equal to zero. Maxwell's equations also do not permit the calculation of the propagation velocity of signals with bandwidth propagating in a lossy medium, and all the published solutions for propagation velocities assume sinusoidal (linear) signals.

In order to remedy this state of affairs, Harmuth proposed an amendment to Maxwell's equations, which I have called the *Harmuth ansatz*.[224–226] The proposed amended equations (in Gaussian form) are

Coulomb's law [Eq. (1.1)]:

$$\nabla \cdot D = 4\pi \rho_e; \tag{5.1.1}$$

Maxwell's generalization of Ampère's law [Eq. (1.2)]:

$$\nabla \times \mathbf{H} = \left(\frac{4\pi}{c}\right) \mathbf{J}_e + \left(\frac{1}{c}\right) \frac{\partial \mathbf{D}}{\partial t}; \tag{5.1.2}$$

Postulate of the presence of free magnetic poles:

$$\nabla \cdot \mathbf{B} = \rho_m; \tag{5.1.3}$$

Faraday's law with magnetic monopole:

$$\nabla \times \mathbf{E} + \left(\frac{1}{c}\right) \frac{\partial \mathbf{B}}{\partial t} + \left(\frac{4\pi}{c}\right) \mathbf{J}_m = 0; \tag{5.1.4}$$

and the constitutive relations:

$$\mathbf{D} = \varepsilon \mathbf{E}, \tag{1.5 and 5.15}$$

$$\mathbf{B} = \mu \mathbf{H}, \tag{1.7 and 5.15}$$

$$\mathbf{J}_e = \sigma \mathbf{E} \text{ (electric Ohm's law)}, \tag{1.6 and 5.15}$$

$$\mathbf{J}_m = s \mathbf{E} \text{ (magnetic Ohm's law)}, \tag{5.1.8}$$

where $\mathbf{J}_e$ is electric current density, $\mathbf{J}_m$ is magnetic current density, ρ_e is electric charge density, ρ_m is magnetic charge density, σ is electric conductivity and s is magnetic conductivity.

Setting $\rho_e = \rho_m = \nabla \cdot \mathbf{D} = \nabla \cdot \mathbf{B} = 0$ for free space propagation gives

$$\nabla \times \mathbf{H} = \sigma \mathbf{E} + \frac{\varepsilon \partial \mathbf{E}}{\partial t}, \tag{5.1.9}$$

$$\nabla \times \mathbf{E} + \frac{\mu \partial \mathbf{H}}{\partial t} + s\mathbf{H} = 0 \tag{5.1.10}$$

$$\varepsilon \nabla \cdot \mathbf{B} = \mu \nabla \cdot \mathbf{B} = 0, \tag{5.1.11}$$

and the equations of motion

$$\frac{\partial E}{\partial y} + \frac{\mu \partial H}{\partial t} + sH = 0, \tag{5.1.12}$$

$$\frac{\partial H}{\partial y} + \frac{\varepsilon \partial E}{\partial t} + \sigma E = 0. \tag{5.1.13}$$

Differentiating Eqs. (5.1.12) and (5.1.13) with respect to y and t permits the elimination of the magnetic field, resulting in [Ref. 212, Eq. (21)]

$$\frac{\partial^2 E}{\partial y^2} - \frac{\mu \varepsilon \partial^2 E}{\partial t^2} - (\mu \sigma + \varepsilon s)\frac{\partial E}{\partial t} - s\sigma E = 0, \tag{5.1.14}$$

which is a two-dimensional Klein–Gordon equation (without boundary conditions) in the sine-Gordon form

$$\frac{\partial^2 E}{\partial y^2} - \left(\frac{1}{c^2}\right)\frac{\partial^2 E}{\partial t^2} - \alpha \sin(\beta E(y, t)) = 0, \tag{5.1.15}$$

$$\alpha \sin(\beta E(y, t)) \approx -\left[\frac{\alpha \beta E}{\partial t} - O(E)\right], \tag{5.1.16}$$

$$\alpha = \exp(+\mu\sigma) \tag{5.1.17}$$

$$\beta = \exp(+\varepsilon\sigma)\beta, \tag{5.1.18}$$

where $(\partial^2 E/\partial y^2 - (1/c^2)\partial^2 E/\partial t^2)$ is the "nonlinear" term and $(\alpha \sin(\beta E(y, t)))$ is the dispersion term. This match of "nonlinearity" and dispersion permits soliton solutions, and the field described by Eq. (5.1.15) has a "mass," $m = \sqrt{\alpha\beta}$.

Equation (5.1.15) may be derived from the Lagrangian density

$$L = \frac{1}{2}\left[\left(\frac{\partial E}{\partial y}\right)^2 - \left(\frac{\partial E}{\partial t}\right)^2\right] - V(E), \tag{5.1.19}$$

where

$$V(E) = \frac{\alpha}{\beta}(1 - \cos \beta E). \tag{5.1.20}$$

The wave equation for E has a solution which can be written in the form

$$E(y, t) = E_E(y, t) = E_0[w(y, t) + F(y)], \tag{5.1.21}$$

where $F(y)$ indicates that an electric step function is the excitation. A wave equation for $F(y)$ is

$$\frac{d^2 F}{dy^2} - s\sigma F = 0, \tag{5.1.22}$$

with the solution

$$F(y) = A_{00}\exp[-yL] + A_{01}\exp\left[\frac{y}{L}\right], \quad L = (s\sigma)^{-1/2}. \tag{5.1.23}$$

Boundary conditions require $A_{01} = 0$ and $A_{00} = 1$, therefore

$$F(y) = \exp\left[-\frac{y}{L}\right]. \tag{5.1.24}$$

Insertion of Eq. (5.1.21) into Eq. (5.1.14) gives [Ref. 212, Eq. (40)]

$$\frac{\partial^2 w}{\partial y^2} - \frac{\mu\varepsilon\partial^2 w}{\partial t^2} - (\mu\sigma + \varepsilon s)\frac{\partial w}{\partial t} - s\sigma w = 0, \tag{5.1.25}$$

which we can again put into sine-Gordon form

$$\frac{\partial^2 w}{\partial y^2} - \left(\frac{1}{c^2}\right)\frac{\partial^2 w}{\partial t^2} - \alpha \sin(\beta w(y, t) = 0. \tag{5.1.26}$$

Harmuth[212] developed a solution to (5.1.21) by seeking a general solution of $w(y, t)$ using a separation-of-variables method (and setting s to zero after a solution is found). This solution works well, but we now indicate another solution. The solutions to the sine-Gordon equation (5.1.15) are the hyperbolic tangents

$$E(y) = \pm \left(\frac{8\sqrt{\alpha\beta}}{\beta^2}\right) \tan^{-1} \left(\exp\left[\frac{y - y_0 - ct}{\sqrt{1 - c^2}}\right]\right), \qquad (5.1.27)$$

where $c = \sqrt{1/\mu\varepsilon}$, which describe solitons. It is also well known that the sine-Gordon and Thirring[256]models are equivalent[145] and that both admit two currents: one a Noether current and the other a topological current.

The following remarks may now be made: the introduction of $F(y) = \exp[-y/L]$, according to the Harmuth ansatz (Ref. 212, p. 253), provides integrability. It is well known that soliton solutions require complete integrability. According to the present view, $F(y)$ also provides the problem with boundary conditions, the *necessary condition* for A_μ potential invariance. Equation (5.1.24) is, in fact, a phase factor [Eqs. (4.1), (4.3), (4.4) and (4.6)]. Furthermore, Eq. (5.1.21) is of the form of Eq. (4.11). Therefore the Harmuth ansatz amounts to a definition of boundary conditions — i.e. obtains the condition of separate electromagnetic field comparison by overlapping fields — *which permits complete integrability and soliton solutions to the extended Maxwell equations*. Furthermore, it was already seen, above, that with boundary conditions defined, the A_μ potentials are gauge-invariant, implying a magnetic monopole and charge. It is also known that the magnetic monopole and charge constructs only exist under certain field symmetries. In the next section, methods are presented for conditioning fields into those higher order symmetries.

5.2. Conditioning the electromagnetic field into altered symmetry: Stokes' interferometers and Lie algebras

The theory of Lie algebras offers a convenient summary of the interaction of the A_μ potential operators with the E fields.[257,258] The relevant parts of the theory are as follows. A manifold, L, is a set of elements in one-to-one correspondence with the points of a vector manifold M. M is a set of vectors called points of M. A Lie group, L, is a group which is also a manifold on which the group operations are continuous. There exists an invertible function, T, which maps each point

x in M to a group element $X = T(x)$ in L. The group M is a global parametrization of the group L.

If $\partial = \partial_x$ is the derivative with respect to a point on a manifold M, then the *Lie bracket* is

$$[a, b] = a \cdot \partial b - b \cdot \partial a = a\nabla \cdot b - b \cdot \nabla a, \qquad (5.2.1)$$

where a and b are arbitrary vector-valued functions. Furthermore, with $\wedge$ signifying the outer product,[227,228]

$$[a, b] = \partial \cdot (a \wedge b) - b\partial \cdot a + a\partial \cdot b, \qquad (5.2.2)$$

showing that the Lie bracket preserves tangency.

The fundamental theorem of Lie group theory is that the Lie bracket $[a, b]$ of differential fields on any manifold is again a vector field. A set of vector fields, $a, b, c, \ldots$, on any manifold forms a Lie algebra if it is closed under the Lie bracket and all fields satisfy the Jacobi identity:

$$[[a, b], c] + [[b, c], a] + [[c, a], b] = 0. \qquad (5.2.3)$$

If $c = 0$, then

$$[a, b] = 0. \qquad (5.2.4)$$

The A_μ potentials effect mappings, T_1, from the global field to the E local fields, considered as group elements in L; and there must be a second mapping, T_2, of those separately conditioned E fields now considered global, onto a single local field for the T_1 mappings to be detected (measurable). That is, in the AB situation (and substituting fields for electrons), if the E fields traversing the two paths are E_1 and E_2, and those fields before and after interaction with the A_μ field are E_{1i} and E_{1f} and E_{2i} and E_{2f} respectively, then $E_{1f}+E_{2f} = T(E_{1i}+E_{2i})$, where $x_1 = E_{1i}$ and $x_2 = E_{2i}$ are points in M, and $X = (E_{1f} + E_{2f})$ is considered a *group* point in L and $T = T_1 + T_2, T_1 = T_1^{-1}$. In the same situation, although $(E_{1f} - E_{2f}) = \exp(i\hbar/e) \oint A_\mu dx^\mu = \Phi$ [i.e. the phase factor detected at Z in a separate second mapping, T_2, in Fig. 5.2.1 can be ascribed to a nonintegrable (path-independent) phase factor], the influence of the first, T_1, mapping or conditioning of $E_{1i} + E_{2i}$ by the A_μ operators along the separate path trajectories *preceded* that second mapping, T_2, at Z. Therefore the A_μ potential

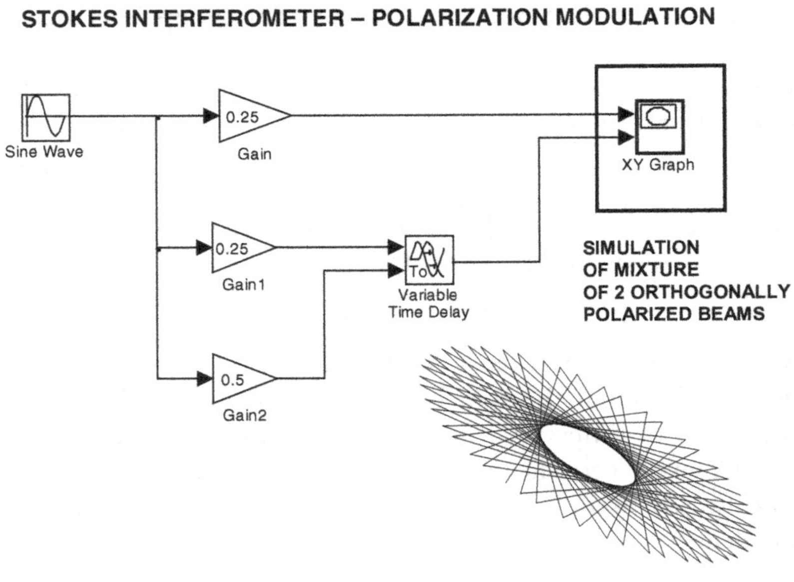

STOKES INTERFEROMETER – POLARIZATION MODULATION

Fig. 5.2.1. Wave guide system paradigm for polarization-modulated ($\partial\phi/\partial t$) wave emission. This is a completely adiabatic system in which oscillating energy enters from the left and exits from the right. On entering from the left, energy is divided into two parts equally. One part, of amplitude $E/2$, is polarization-rotated and used in providing phase modulation, $\partial\phi/\partial t$ — this energy is spent (absorbed) by the system in achieving the phase modulation; the other part, of amplitude $E/2$, is divided into two parts equally, so that two oscillating wave forms of amplitude $E/4$ are formed for later superposition at the output. Owing to the phase modulation of one of them with respect to the other, $0 < \phi < 360°$, and their initial orthogonal polarization, the output is of continuously varying polarization. The choice of wave division into two equal parts is arbitrary. (From Refs. 230 and 231.)

field operators produce a mapping of the global spatiotemporal conditions onto local fields, which, in the case we are considering, are the separate $\mathbf{E}_{1i} + \mathbf{E}_{2i}$ fields. Thus, according to this conception, the $\mathbf{A}_\mu$ potentials are local operator fields mapping the local-to-global gauge $(T_1 : M \to L)$, whose effects are detectable at a later spatiotemporal position only at an overlapping (X group) point, i.e. by a second mapping $(T_2 : L \to M)$, permitting comparisons of the differently conditioned fields in a many-to-one (global-to-local) summation.

If $a = \mathbf{E}_{1i}$ and $b = \mathbf{E}_{2i}$, where $\mathbf{E}_{1i}$ and $\mathbf{E}_{2i}$ are local field intensities and $c = \mathbf{A}_\mu$, i.e. $\mathbf{A}_\mu$ is a local field mapping $(T_1 : M \to L)$ according to gauge conditions specified by boundary conditions, then the field interactions of a, b and c, or $\mathbf{E}_{1i}$, $\mathbf{E}_{2i}$ and $\mathbf{A}_\mu$, are described by the Jacobi identity [Eq. (5.2.3)]. If $c = \mathbf{A}_\mu = 0$, then $[a, b] = 0$. With the Lorentz gauge (or boundary conditions), the $\mathbf{E}_{1i}$, $\mathbf{E}_{1f}$, $\mathbf{E}_{2i}$ and $\mathbf{E}_{2f}$ field relations are described by SU(2) symmetry. With other boundary conditions and no separate $\mathbf{A}_\mu$ conditioning, the $\mathbf{E}_{1i}$ and $\mathbf{E}_{2i}$ fields (there are no $\mathbf{E}_{1f}$ and $\mathbf{E}_{2f}$ fields in this situation) are described by U(1) symmetry relations.

The T_1, T_2 mappings can be described by classical control theory analysis and the $\mathbf{A}_\mu$ potential conditioning can be given a physical wave guide interferometer representation (cf. Ref. 229). The wave guide system considered here is completely general in that the output can be phase-, frequency- and amplitude-modulated. It is an adiabatic system (lossless) and only three of the lines are wave guides — the input, the periodically delayed line, and the output. Other lines shown are energy-expending, phase-modulating lines. The basic design is shown in Fig. 5.2.1. In this figure, the input is $E = E \exp(i\omega t)$. The output is

$$E_{\text{OUT}} = \left(\frac{E}{4}\right) \exp(i\omega t) + \left(\frac{E}{4}\right) \exp(i[\omega + \exp(i\omega t) - 1]t), \quad (5.2.5)$$

where, referring to Fig 5.2.1, $\phi = F(E/2)$ and $\partial\phi/\partial t = F'(E/2)$, and it is understood that the first arm is orthogonal to the second.

The wave guide consists of two arms — the upper $(E/4)$ and the second $(E/4)$, with which the upper is combined. The lower, or third arm, merely expends energy in achieving the phase modulation of the second arm with respect to the first. This can be achieved by merely making the length of the second arm change in a sinusoidal fashion (i.e. producing a $\partial\phi/\partial t$ with respect to the first arm), or it can be achieved electro-optically. Whichever way is used, one half the total energy of the system $(E/2)$ is spent on achieving the phase modulation in the particular example shown in Fig. 5.2.1. The entropy change from input to output of the wave guide is compensated for by energy expenditure in achieving the phase modulation which causes the entropy change.

One can nest phase modulations. The next order nesting is shown in Fig. 5.2.2, and other, higher order nestings of order n, for the cases

$\partial \phi^n / \partial t^n$, $n = 2, 3, \ldots$, follow the same procedure. The input is again $E = E \exp(i\omega t)$. The output is

$$E_{\text{OUT}(n=2)} = \left(\frac{E}{4}\right) \exp(i\omega t) + \left(\frac{E}{4}\right) \exp(i[\phi_1 + \exp(i\phi_2 t) - 1]t),$$

$$(5.2.6)$$

where $\phi_1 = F_1(E/4)$, $\phi_2 = F_2(E/4)$ and $\partial \phi^2 / \partial t^2 = F_1^2 \cdot F_2^2$, and again it is understood that the first arm is orthogonal to the second.

Again, the wave guide consists of two arms — the upper $(E/4)$ and the second $(E/4)$, with which the upper is recombined. The lower two arms, three and four, merely expend energy in achieving the phase modulation of the second arm with respect to the first. This again can be achieved by merely making the length of the second arm change in a sinusoidal fashion (i.e. producing a $\partial \phi^2 / \partial t^2$ with respect to the first arm), or it can be achieved electro-optically for visible frequencies. Whichever way is used, one half the total energy of the system $(E/4 + E/4 = E/2)$ is spent on achieving the phase modulation of the particular sample shown in Fig. 5.2.2.

Both the systems shown in Figs. 5.2.1 and 5.2.2, and all higher orders of such systems, $\partial \phi^n / \partial t^n$, $n = 1, 2, 3, \ldots$, are adiabatic with respect to the total field, and Poynting's theorem applies to them all. However, the Poynting description, or rather limiting condition, is insufficient to describe these fields exactly, and neglects the orthogonal polarization two-beam picture, and a more exact analysis is provided by the control theory picture shown here.

These wave guides we shall call Stokes interferometers. The Stokes equation is [Eq. (3.3.1)]

$$\oint \mathbf{A} \cdot dl = \int_S (\nabla \times \mathbf{A}) \cdot \mathbf{n} \, da, \qquad (5.2.7)$$

and the energy-expending lines of the two Stokes interferometers shown are normal to the two wave guide lines. l is varied sinusoidally,

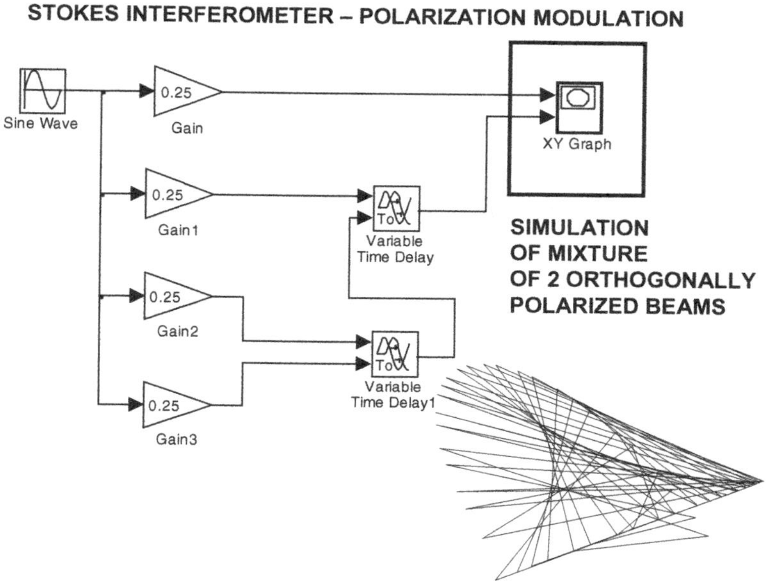

STOKES INTERFEROMETER – POLARIZATION MODULATION

Fig. 5.2.2. Wave guide system paradigm for polarization-modulated $(\partial\phi^2/\partial t^2)$ wave emission. This is a completely adiabatic system in which oscillating energy enters from the left and exits from the right. On entering from the left, the energy is divided into two parts equally. One part, of amplitude $E/2$, is used in providing phase modulation, $\partial\phi^2/\partial t^2$ — this energy is spent (absorbed) by the system in obtaining the phase modulation; the other part, of amplitude $E/2$, is divided into two parts equally, one of which is polarization-rotated, so that two oscillating wave forms of amplitude $E/4$ but initially orthogonally polarized are formed for later superposition at the output. Unlike the system shown in Fig 5.2.1, the energy expended on phase-modulating one of these waves is divided into two parts equally, of amplitude $E/4$, one of which is phase-modulated, $\partial\phi/\partial t$, with respect to the other, as in Fig. 5.2.1. The energy of the superposition of these two waves is then expended to provide a second phase-modulated $\partial\phi^2/\partial t^2$ wave which is superposed with the nondelayed wave. Owing to the phase modulation of one of them with respect to the other, $0 < \phi < 360°$, and their initial orthogonal polarization, the output is of continuously varying polarization. The choice of wave division into two equal parts is arbitrary. (From Refs. 230 and 231.)

so we have

$$\oint A \sin \omega t \, dl = \int_S (\nabla \times \mathbf{A}) \cdot \mathbf{n} \, da = E_{\text{OUT}\,n=1} \quad \text{(Fig. 5.2.1)} \quad (5.2.8)$$

$$\oint A \sin \omega t \, dl = \int_S (\nabla \times \mathbf{A}) \cdot \mathbf{n} \, da = E_{\text{OUT}\,n=2} \quad \text{(Fig. 5.2.2)} \quad (5.2.9)$$

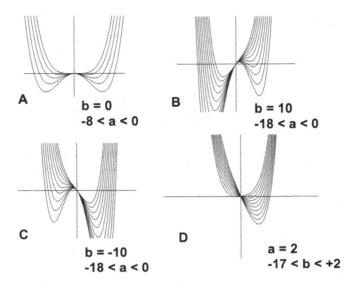

Fig. 5.2.3.   Plots of $A = 1/4x^4 + 1/2ax^2 + bx + c$. (A) $b = 0$ and various values of a; (B) $b = 10$ and various values of a; (C) $b = -10$ and various values of a; (D) $a = 2$ and various values of b. For positive values of a, SU(2) symmetry is restored. For negative values of a, symmetry is broken and U(1) symmetry is obtained. (From Ref. 232.)

The gauge symmetry consequences of this conditioning are shown in Figs. 5.2.3 and 5.2.4. The potential, $\mathbf{A}_\mu$, in Taylor expansion along one coordinate is

$$A = \frac{1}{4}x^4 + \frac{1}{2}ax^2 + bx + c, \tag{5.2.10}$$

with $b < 0$ in the case of E_{in} and $b > 0$ in the case of E_{OUT}. A Stokes interferometer permits the E field to restore a symmetry which was broken before this conditioning. Thus the E_{in} field is in U(1) symmetry form and the E_{OUT} field is conditioned to be in SU(2) symmetry form. The conditioning of the **E** field to SU(2) symmetry form is the opposite of symmetry breaking. It is well known that the Maxwell theory is in U(1) symmetry form and the theoretical constructs of the magnetic monopole and charge exist in SU(2) symmetry form.[225,226,230–234]

Other interferometric methods besides Stokes interferometer polarization modulators which restore symmetry are cavity wave

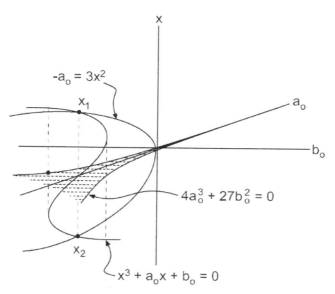

Fig. 5.2.4. A representative system defined in the (x, a, b) space. Other systems can be represented in the cusp area at other values of x, a and b. As in Fig. 5.2.3, for positive values of a, SU(2) symmetry is restored. For negative values of a, symmetry is broken and U(1) symmetry is obtained. (From Ref. 232.)

guide interferometers. For example, the Mach–Zehnder and Fabry–Perot interferometers are SU(2) conditioning interferometers[235] (Fig. 5.2.5). The SU(2) group characterizes passive lossless devices with two inputs and two outputs with the boson commutation relations

$$[E_{1*}, E_{2*}] = [E_{1*}^{\Uparrow}, E_{2*}^{\Uparrow}] = 0, \qquad (5.2.11)$$

$$[E_{1*}, E_{2*}^{\Uparrow}] = d_{12*}, \qquad (5.2.12)$$

where $E^{\Uparrow}$ is the Hermitian conjugate of E and $*$ signifies both *in* (entering) and *out* (exiting) fields, i.e. before and after $\mathbf{A}_\mu$ conditioning. The Hermitian operators are

$$\mathbf{J}_x = \frac{1}{2}(E_1^{\Uparrow} E_{2*} + E_{1*} E_2^{\Uparrow}) = \frac{1}{2}(A_1 \times B_{1\text{IN}} + B_{2\text{IN}} \times A_2)$$

$$= (\mathbf{A} \times \mathbf{B} - \mathbf{B} \times \mathbf{A}), \qquad (5.2.13a)$$

A. FABRY-PEROT INTERFEROMETER

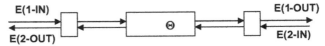

B. MACH-ZEHNDER INTERFEROMETER

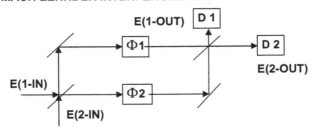

C. STOKES

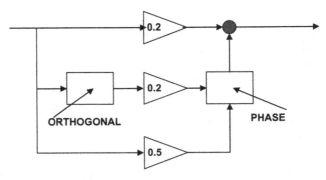

Fig. 5.2.5. SU(2) field conditioning interferometers: (A) Fabry–Perot; (B) Mach-Zehnder; (C) Stokes. (After Ref. 225.)

$$\mathbf{J}_y = -\frac{i}{2}(E_1^{\Uparrow} E_{2*} - E_{1*} E_2^{\Uparrow}) = -\frac{i}{2}(A_1 \times B_{1\text{IN}} + B_{2\text{IN}} \times A_2)$$

$$= (\mathbf{A} \cdot \mathbf{B} - \mathbf{B} \cdot \mathbf{A}), \qquad\qquad (5.2.13\text{b})$$

$$\mathbf{J}_z = \frac{1}{2}(E_1^{\Uparrow} E_{1*} - E_{2*} E_2^{\Uparrow}) = \frac{1}{2}(A_1 \times E_{1\text{OUT}} - E_{2\text{OUT}} \times A_2)$$

$$= (\mathbf{A} \times \mathbf{E} - \mathbf{E} \times \mathbf{A}), \qquad\qquad (5.2.13\text{c})$$

$$i\mathbf{J}_z = \frac{1}{2}(E_1^{\Uparrow} E_{1*} + E_{2*} E_2^{\Uparrow}) = -\frac{i}{2}(A_1 \times E_{1\text{OUT}} + E_{2\text{OUT}} \times A_2)$$

$$= (\mathbf{A} \times \mathbf{E} + \mathbf{E} \times \mathbf{A}), \qquad\qquad (5.2.13\text{d})$$

where the substitutions are

$$E_{1*} = B_{2\text{IN}} \text{ and } \times E_{1\text{OUT}},$$
$$E_{2*} = \times B_{1\text{IN}} \text{ and } E_{2\text{OUT}},$$
$$E_1^{\Uparrow} = A_1,$$
$$E_2^{\Uparrow} = \times A_2,$$

(5.2.14)

satisfying the Lie algebra

$$[J_x, J_y] = iJ_z,$$
$$[J_y, J_z] = iJ_x,$$
$$[J_z, J_x] = iJ_y.$$

(5.2.15)

The analysis presented in this section is based on the relation of induced angular momentum to the *eduction* of gauge invariance (see also Ref. 236). One gauge-invariant quantity or observable in one gauge or symmetry can be covariant with another in another gauge or symmetry. The Wu–Yang condition of field overlap, permitting measurement of $\Phi = i \oint \exp \mathbf{A}_\mu d^\mu$, requires coherent overlap. All other effects are either observed at low temperature where thermodynamic conditions provide coherence, or are a self-mapping, which also provides coherence. Thus the question of whether classical $\mathbf{A}_\mu$ wave effects can be observed at long range reduces to the question of how far the coherence of the two fields can be maintained.

Recently, Oh *et al.*[237] have derived the nonrelativistic propagator for the generalized AB effect, which is valid for any gauge group in a general multiply connected manifold, as a gauge artifact in the universal covering space. These authors conclude that (1) if a partial propagator along a multiply connected space (M in the present notation) is lifted to the universal covering space (L in the present notation), i.e. $T_1: M \to L$, then (2) for a gauge transformation $U(x)$ of $\mathbf{A}_\mu$ on the covering space L, an AB effect will arise if (3) $U(x)$ is not projectable to be a well-defined single-valued gauge transformation on M, but (4) $\mathbf{A}_\mu = U(x)\partial_\mu U(x)^{-1}$ (i.e. $T_1 T_2^{-1}$) is neverthelesss projectable, i.e. for a $T_2 : L \to M$, in agreement with the analysis presented here. We have stressed, however, that the $\mathbf{A}_\mu = T_1 : L \to M$ have a physical existence, whether the $T_2 : L \to M$ mapping exists or can be performed or not. Naturally, if this second mapping is not performed, then no AB effect exists (i.e. no comparative mapping exists).

Although interferometric methods can condition fields into SU(2) or other symmetric form, there is, of course, no control over the space–time metric in which those fields exist. When the conditioned field leaves the interferometer, at time $t = 0$, the field is in exact SU(2) form. At time $t > 0$, the field will depart from SU(2) form in as much as it is scattered or absorbed by the medium.

The gauge invariance of the phase factor requires a multiply connected field. In the case of quantum particles, this would mean wave function overlap of two individual quanta. Classically, however, *every* polarized wave is made us of two polarized vectorial components. Therefore, classically, every polarized wave is a multiply connected field (cf. Ref. 238) in SU(2) form. However, the extension of the Maxwell theory to SU(2) form, i.e. to non-Abelian Maxwell theory, defines mutiply connected local fields in a global covering space, i.e. in a global simply connected form. In the next section, the Maxwell equations redefined in SU(2)/Z_2, non-Abelian, or multiply connected form are examined.

5.3. *Non-Abelian Maxwell equations*

Using Yang–Mills theory,[27] the non-Abelian Maxwell equations, which describe SU(2)-symmetry-conditioned radiation, become Coulomb's law:

$$\text{no existence in SU(2) symmetry;} \tag{5.3.1}$$

Ampère's law:

$$\frac{\partial \mathbf{E}}{\partial t} - \nabla \times \mathbf{B} + iq[A_0, \mathbf{E}] - iq(\mathbf{A} \times \mathbf{B} - \mathbf{B} \times \mathbf{A}) = -\mathbf{J}, \tag{5.3.2}$$

the presence of free "magnetic monopoles" (instantons):

$$\nabla \cdot \mathbf{B} + iq(\mathbf{A} \cdot \mathbf{B} - \mathbf{B} \cdot \mathbf{A}) = 0; \tag{5.3.3}$$

Faraday's law:

$$\nabla \times \mathbf{E} + \frac{\partial \mathbf{B}}{\partial t} + iq[A_0, \mathbf{B}] + iq(\mathbf{A} \times \mathbf{E} - \mathbf{E} \times \mathbf{A}) = 0; \tag{5.3.4}$$

and the current relation

$$\nabla \cdot \mathbf{E} - \mathbf{J}_0 + iq(\mathbf{A} \cdot \mathbf{E} - \mathbf{E} \cdot \mathbf{A}) = 0. \tag{5.3.5}$$

Coulomb's law [Eq. (5.3.1)] amounts to an imposition of spherical symmetry requirements, as a single isolated source charge permits the choice of charge vector to be arbitrary at every point in space–time. Imposition of this symmetry reduces the non-Abelian Maxwell equations to the same form as conventional electrodynamics, i.e. to Abelian form.

Harmuth's ansatz is the addition of a magnetic current density to Maxwell's equations — an addition which may be set to zero after completion of calculations.[224] With a magnetic current density, Maxwell's equations describe a space–time field of higher order symmetry and consist of invariant physical quantities (e.g. the field $\partial_x \mathbf{F} = J$), magnetic monopole and charge. Harmuth's amended equations are [Ref. 212, Eqs. (4)–(7)]

$$\nabla \times \mathbf{H} = \frac{\partial \mathbf{D}}{\partial t} + g_e, \tag{5.3.6a}$$

$$-\nabla \times \mathbf{E} = \frac{\partial \mathbf{B}}{\partial t} + g_m, \tag{5.3.6b}$$

$$\nabla \cdot \mathbf{D} = \rho_e, \tag{5.3.6c}$$

$$\nabla \cdot \mathbf{B} = \rho_m, \tag{5.3.6d}$$

$$g_e = \sigma \mathbf{E}, \tag{5.3.6e}$$

$$g_m = s\mathbf{H}, \tag{5.3.6f}$$

where g_e, g_m, ρ_e, ρ_m and s are electric current density, magnetic current density, electric charge density, magnetic charge density and magnetic conductivity, respectively.

It should be noted that classical magnetic sources or instanton-like sources are not dismissed by all researchers in classical field theory. For example, Tellegren's formulation for the gyrator[239] and formulations needed in descriptions of chiral media (natural optical activity)[240,241] require magnetic source terms even in the frequency domain. The ferromagnetic aerosol experiments by Mikhailov on the Ehrenhaft effect also imply a magnetic monopole instanton or pseudoparticle interpretation,[242–246] the spherical symmetry of the aerosol particles providing SU(2) boundary conditions according to the present view.

Comparing the SU(2) formulation of the Maxwell equations and the Harmuth equations reveals the following identities[224]

U(1) Symmetry: SU(2) Symmetry:

$$\rho_e = J_0 \qquad \rho_e = J_0 - iq(\mathbf{A} \cdot \mathbf{E} - \mathbf{E} \cdot \mathbf{A}) = J_0 + qJ_z \qquad (5.3.7a)$$

$$\rho_m = 0 \qquad \rho_m = -iq(\mathbf{A} \cdot \mathbf{B} - \mathbf{B} \cdot \mathbf{A}) = -qJ_y \qquad (5.3.7b)$$

$$g_e = \mathbf{J} \qquad g_e = iq[A_0, \mathbf{E}] - iq(\mathbf{A} \times \mathbf{B} - \mathbf{B} \times \mathbf{A}) + \mathbf{J}$$

$$\qquad\qquad = iq[A_0, \mathbf{E}] - iqJ_x + \mathbf{J} \qquad (5.3.7c)$$

$$g_m = 0 \qquad g_m = iq[A_0, \mathbf{B}] - iq(\mathbf{A} \times \mathbf{E} - \mathbf{E} \times \mathbf{A})$$

$$\qquad\qquad = iq[A_0, \mathbf{B}] - iqJ_z \qquad (5.3.7d)$$

$$\sigma = \frac{\mathbf{J}}{\mathbf{E}} \qquad \sigma = \{iq[A_0, \mathbf{E}] - iq(\mathbf{A} \times \mathbf{B} - \mathbf{B} \times \mathbf{A}) + \mathbf{J}\}/\mathbf{E}$$

$$\qquad\qquad = \{iq[A_0, \mathbf{E}] - iqJ_x + \mathbf{J}\}/\mathbf{E} \qquad (5.3.7e)$$

$$s = 0 \qquad s = \{iq[A_0, \mathbf{B}] - iq(\mathbf{A} \times \mathbf{E} - \mathbf{E} \times \mathbf{A})\}/\mathbf{H}$$

$$\qquad\qquad = \{iq[A_0, \mathbf{B}] - iqJ_z\}/\mathbf{H}. \qquad (5.3.7f)$$

It is well known that only some topological charges are conserved (i.e. are gauge-invariant) after symmetry breaking — electric charge is, magnetic charge is not.[191] Therefore, the Harmuth ansatz of setting magnetic conductivity [and other SU(2) symmetry constructs] to zero on conclusion of signal velocity calculations has a theoretical justification. It is also well known that some physical constructs which exist in both a lower and a higher symmetry form are more easily calculated for the higher symmetry, transforming to the lower symmetry after the calculation is complete. The observables of the electromagnetic field exist in a U(1) symmetry field. Therefore the problem is to relate invariant physical quantities to the variables employed by a particular observer. This means a mapping of space–time vectors into space vectors, i.e. a space–time split.

This mapping is not necessary for solving and analyzing the basic equations. As a rule, it only complicates the equations needlessly.

Therefore, the appropriate time for a split is usually after the equations have been solved. It is appropriate to mention here the interpretation of the AB effect offered by Bernido and Inomata.[57] These authors point out that a path integral can be explicitly formulated as a sum of partial propagators corresponding to homotopically different paths. In the case of the AB effect, the mathematical object to be computed in this approach is a *propagator* expressed as a path integral in the *covering space* of the background physical space. Therefore, the path dependence of the AB phase factor is wholly of topological origin and the AB problem is reduced to showing that the full propagator can be expressed as a sum of partial propagators belonging to all topological inequivalent paths. The paths are partitioned into their homotopy equivalence classes, Feynman sums over paths in each class giving homotopy propagators and the whole effect of the gauge potential being to multiply these homotopy propagators by different gauge phase factors. However, the relevant point, with respect to the Harmuth ansatz is that the full propagator is expressed in terms of the covering space, rather than the physical space. The homotopy propagators are related to propagators in the universal covering manifold, leading to an expansion of the propagators in terms of eigenfunctions of a Hamiltonian on the covering manifold.

The approach to multiply connected spaces offered by Dowker[247] and Sundrum and Tassic[59] also uses the covering space concept. A multiply connected space, M, and a universal covering space, M^*, are defined

$$M^* \to M = \frac{M^*}{G}, \tag{5.3.7}$$

where G is a properly continuous, discrete group of isometries of M^*, without fixed points, and M^* is simply connected. Each group of M corresponds to n different points qg of M^*, where g ranges over the n elements of G. M^* is then divided into subsets of a finite number of points or fibers, one fiber corresponding to one point of M. M^* is a bundle or fibered space, and Γ is the group of the bundle. The major point, in the present instance, is that the propagator is given in terms of a matrix representation of the covering space M^*. Harmuth calculates the propagation in the covering space where the Hamiltonian is self-adjoint. Self-adjointness means that non-Hermitian components are compensated for.[248] Thus, the propagation in the covering space is well defined.

Consequently, Harmuth's ansatz can be interpreted as: (i) a mapping of Maxwell's (U(1) symmetrical) equations into a higher order symmetry field [of SU(2) symmetry] — a symmetry which permits the definition of magnetic monopoles (instantons) and magnetic charge; (ii) solving the equations for propagation velocities; and (iii) mapping the solved equations back into the U(1) symmetrical field (thereby removing the magnetic monopole and charge constructs).

6. Discussion

The concept of the electromagnetic field was formed by Faraday and set in a mathematical frame by Maxwell to describe electromagnetic effects in a space–time region. It is a concept addressing *local* effects. The Faraday–Maxwell theory, which was founded on the concept of the electrotonic state, *potentially* had the capacity to describe global effects but the manifestation of the electrotonic state, the **A** field, was abandoned in the later interpretation of Maxwell. When, in this interpretation, the theory was refounded on the field concept, and the issue of energy propagation was examined, action-at-a-distance (Newton) was replaced by contact-action (Descartes). That is, a theory (Newton's) accounting for both local and global effects was replaced by a completely local theory (Descartes'). The contemporary local theory can address global effects with the aid of the Lorentz invariance condition, or Lorentz gauge. However, Lorentz invariance is due to a chosen gauge and chosen boundary conditions, and these are not an inevitable consequence of the (interpreted) Maxwell theory, which became a theory of local effects.

According to the conventional viewpoint, the local field strength, $F_{\mu\nu}$, completely describes electromagnetism. However, owing to the effects discussed here, there is reason to believe that $F_{\mu\nu}$ does not describe electromagnetism completely. In particular, it does not describe *global* effects resulting in different histories of local spatiotemporal conditioning of the constituent parts of summed multiple fields.

Weyl[23-26] first proposed that the electromagnetic field can be formulated in terms of an Abelian gauge transformation. But the Abelian gauge only describes local effects. It was Yang and Mills[27] who extended the idea to non-Abelian groups. The concepts of the Abelian electromagnetic field — electric charge, **E** and **H** fields — are explained within the context of the non-Abelian concepts of magnetic charge and monopole. The Yang–Mills theory *is* applicable to both local and global effects.

If the unbroken gauge group is non-Abelian, only some of the topological charges are gauge-invariant. The electric charge is, the magnetic charge is not.[191] That is the reason magnetic sources are not seen in Abelian Maxwell theory which has boundary conditions which do not compactify or reconstitute symmetry and degrees of freedom.

The $\mathbf{A}_\mu$ potentials have an ontology or physical meaning as *local* operators mapping onto *global* spatiotemporal conditions the *local* e.m. fields. This operation is measurable if there is a second comparative mapping of the conditioned local fields in a many-to-one fashion (multiple connection). In the case of a single local (electromagnetic) field, this second mapping is ruled out — but such an isolated local field is only imaginary, because the imposition of boundary conditions implies the existence of separate local conditions and thereby always a global condition. Therefore, practically speaking, the $\mathbf{A}_\mu$ potentials always have a gauge-invariant physical existence. The $\mathbf{A}_\mu$ potential gauge invariance implies the theoretical constructs of a magnetic monopole (instanton) and magnetic charge, but with no singularities. These latter constructs are, however, confined to SU(2) field conditioning, whereas the $\mathbf{A}_\mu$ potentials have an existence in *both* U(1) *and* SU(2) symmetries.

The physical effects of the $\mathbf{A}_\mu$ potentials are observable empirically at the quantum level (effects 1–5) and at the classical level (2, 3 and 6). The Maxwell theory of fields, restricted to a description of *local intensity* fields, and with the SU(2) symmetry broken to U(1), requires no amendment at all. If, however, the intention is to describe both local *and* global electromagnetism, then an amended Maxwell theory is required in order to include the local operator field of the A_μ potentials, the integration of which describes the phase relations between local intensity fields of different spatiotemporal history after global-to-local mapping.

With only the constitutive relations of e.m. fields to matter defined (and not those of fields to vacuum), contemporary opinion is that the dynamic attribute of force resides in the medium-independent fields, i.e., they are fields of force. As the field–vacuum constitutive relations are lacking, this view can be contested, giving rise to competing accounts of where force resides, such as the opposing view of force not residing in the fields but in the matter (cf. Ref. 249).

The uninterpreted Maxwell, of course, had *two types* of constitutive relations in mind, the second one referring to the energy–medium relation: "…whenever energy is transmitted from one body to another in time, there must be a medium or substance in which the energy exists after it leaves one body and before it reaches the other…" (Maxwell,[8] vol. II, p. 493).

After removal of the medium from consideration, only one constitutive relation remained and the fields have continued to exist as the classical limit of quantum-mechanical exchange particles. However, that cannot be a true existence for the classical force fields because those quantum-mechanical particles are in units of action, not force.

References

1. Belavin, A.A., Polyakov, A.M., Schwartz, A.S. & Tyupkin, Y.S., "Pseudoparticle solutions of the Yang–Mills equations," *Phys. Lett.*, Vol. 59B, pp. 85–87, 1975.
2. Atiyah, M.F., *Michael Atiyah: Collected Works — Volume 5, Gauge Theories* (Clarendon Oxford, 1988).
3. Barut, A.O., "$E = \_h\omega$," *Phys. Lett.* Vol. A143, pp. 349–352, 1991.
4. Shaarawi, A.M., Besieris, I.M. & Ziolkowski, R., "A novel approach to the synthesis of nondispersive wave packet solutions to the Klein–Gordon and Dirac equations," *J. Math. Phys.*, Vol. 31, pp. 2511–2519, 1990.

5. Mason, L.J. & Sparling, G.A.J., "Nonlinear Schrödinger and Korteweg–de Vries are reductions of self-dual Yang–Mills," *Phys. Lett.*, Vol. A137, pp. 29–33, 1989.

6. Maxwell, J.C., "On Faraday's lines of force," *Trans. Camb. Phil. Soc.*, Vol. 10, Part I, pp. 27–83, 1856.

7. Maxwell, J.C., "A dynamical theory of the electromagnetic field," *Phil. Trans. Roy. Soc.*, Vol. 155, pp. 459–512, 1865.

8. Maxwell, J.C., *A Treatise on Electricity and Magnetism*, 2 vols., 1st edn., 1873; 2nd edn., 1881; 3rd edn., 1891 (Oxford, Clarendon), republished in two volumes (Dover, New York, 1954).

9. Bork, A.M., "Maxwell and the vector potential," *Isis*, Vol. 58, pp. 210–222, 1967.

10. Bork, A.M., "Maxwell and the electromagnetic wave equation," *Am. J. Phys.*, Vol. 35, pp. 844–849, 1967.

11. Everitt, C.W.F., *James Clerk Maxwell* (Scribner's, New York, 1975).

12. Hunt, B.J., "The Maxwellians," a dissertation submitted to Johns Hopkins University with the requirements for the degree of Doctor of Philosophy (Baltimore, Maryland, 1984).

13. Buchwald, J.Z., *From Maxwell to Microphysics* (University of Chicago Press, 1985).

14. Hendry, J., *James Clerk Maxwell and the Theory of the Electromagnetic Field* (Adam Hilger, Bristol, 1986).

15. O'Hara, J.G. & Pricha, W., *Hertz and the Maxwellians: A Study and Documentation of the Discovery of Electromagnetic Wave Radiation, 1873–1894* (Peter Peregrinus, London, 1987).

16. Nahin, P.J., *Oliver Heaviside: Sage in Solitude* (IEEE Press, New York, 1988).

17. Faraday, M., *Experimental Researches in Electricity*, 3 vols., 1839, 1844, 1855. Taylor & Francis, vols. 1 and 3; Richard & John Edward Taylor, Vol. 2 (Dover, New York, 1965).

18. Woodruff, A.E., "The contributions of Hermann von Helmholtz to electrodynamics," *Isis*, Vol. 59, pp. 300–311, 1968.

19. Feynman, R.P., Leighton, R.B. & Sands, M., *The Feynman Lectures on Physics*, Vol. 1 (Addison–Wesley, Reading, MA, 1964).

20. Chisholm, J.S.R. & Common, A.K., *Clifford Algebras and Their Applications in Mathematical Physics* (D. Reidel, Dordrecht, 1986).

21. Whittaker, E.T., "On the partial differential equations of mathematical physics," *Math. Ann.*, Vol. 57, pp. 333–355, 1903.

22. Whittaker, E.T., "On an expression of the electromagnetic field due to electrons by means of two scalar potential functions," *Proc. Lond. Math. Soc.*, series 2, Vol. 1, pp. 367–372, 1904.

23. Weyl, H., "Gravitation und Elektrizität," *Sitzungsberichte der Königlishe Preussischen Akademie der Wissenschaften zu Berlin*, 465–480, 1918.

24. Weyl, H., "Eine neue Erweiterung der RelativitΣts theorie," *Ann. Physik*, Vol. 59, pp. 101–133, 1919.

25. Weyl, H., *Gruppen Theorie und Quantenmechanik*, S. Hirzel, Leipzig, 1928 (*The Theory of Groups and Quantum Mechanics*; Dover, 1949).

26. Weyl, H., *The Classical Groups, Their Invariants and Representations* (Princeton University Press, 1939).

27. Yang, C.N., & Mills, R.L., "Conservation of isotopic spin and isotopic gauge invariance," *Phys. Rev.*, Vol. 96, pp. 191–195, 1954.

28. Wilczek, F. & Zee, A., "Appearance of gauge structure in simple dynamical systems," *Phys. Rev. Lett.*, Vol. 52, no. 24, 11 Jun., pp. 2111–2114, 1984.

29. Atiyah, M.F., "Instantons in two and four dimensions," *Commun. Math. Phys.*, 93, pp. 437–451, 1984.

30. Atiyah, M., Hitchin, N.J. & Singer, I.M., "Deformations of instantons," *Proc. Nat. Acad. Sci.*, Vol. 74, pp. 2662–2663, 1977.

31. Atiyah, M., Hitchin, N.J., Drinfeld, V.G. & Manin, Y.I., "Construction of instantons," *Phys. Lett.*, Vol. 65A, pp. 185–187, 1978.

32. Atiyah, M. & Ward, R.S., "Instantons and algebraic geometry," *Commun. Math. Phys.*, Vol. 55, pp. 117–124, 1977.

33. Atiyah, M. & Jones, J.D.S., "Topological aspects of Yang–Mills theory," *Commun. Math. Phys.*, Vol. 61, pp. 97–118, 1978.

34. Barrett, T.W., "On the distinction between fields and their metric: The fundamental difference between specifications concerning medium-independent fields and constitutive specifications concerning relations to the medium in which they exist," *Annales de la Fondation Louis de Broglie*, Vol. 14, pp. 37–75, 1989.

35. Chakravarty, S., Ablowitz, M.J. & Clarkson, P.A., "Reductions of self- dual Yang–Mills fields and classical systems," *Phys. Rev. Lett.*, Vol. 65, pp. 1085–1087, 1990.

36. Yang, C.N., "Hermann Weyl's contribution to physics," pp. 7–21, in K. Chandrasekharan (ed.), *Hermann Weyl, 1885–1985* (Springer, New York, 1986).

37. Aharonov, Y. & Bohm, D., "Significance of electromagnetic potentials in the quantum theory," *Phys. Rev.*, Vol. 115, no. 3, Aug. 1, pp. 485–491, 1959.

38. Aharonov, Y. & Bohm, D., "Further considerations on electromagnetic potentials in quantum theory," *Phys. Rev.*, Vol. 123, no 4, Aug. 15, pp. 1511–1524, 1961.

39. Aharonov, Y. & Bohm, D., "Remarks on the possibility of quantum electrodynamics without potentials," *Phys. Rev.*, Vol. 125, no. 6, Mar. 15, pp. 2192–2193, 1962.

40. Aharonov, Y. & Bohm, D.,"Further discussion of the role of electromagnetic potentials in the quantum theory," *Phys. Rev.*, Vol. 130, no. 4, 15 May, pp. 1625–1632, 1963.

41. Chambers, R.G., "Shift of an electron interference pattern by enclosed magnetic flux," *Phys. Rev. Lett.*, Vol. 5, Jul., pp. 3–5, 1960.

42. Boersch, H., Hamisch, H., Wohlleben, D. & Grohmann, K., "Antiparallele Weissche Bereiche als Biprisma fnr Elektroneninterferenzen," *Zeitschrift fnr Physik*, Vol. 159, pp. 397–404, 1960.

43. Möllenstedt, G. & Bayh, W., "Messung der kontinuierlichen Phasenschiebung von Elektronenwellen im kraftfeldfreien Raum durch das magnetische Vektorpotential einer Luftspule," *Die Naturwissenschaften*, Vol. 49, no. 4, Feb, pp. 81–82, 1962.

44. Matteucci, G. & Pozzi, G., "New diffraction experiment on the electrostatic Aharonov–Bohm effect," *Phys. Rev. Lett.*, Vol. 54, no. 23, Jun., pp. 2469–2472, 1985.

45. Tonomura, A., Matsuda, T., Suzuki, R., Fukuhara, A., Osakabe, N., Umezaki, H., Endo, J., Shinagawa, K., Sugita, Y. & Fujiwara, H., "Observation of Aharonov–Bohm effect by electron microscopy," *Phys. Rev. Lett.*, Vol. 48, pp. 1443–1446, 1982.

46. Tonomura, A., Umezaki, H., Matsuda, T., Osakabe, N., Endo, J. & Sugita, Y., "Is magnetic flux quantized in a toroidal ferromagnet?", *Phys. Rev. Lett.*, Vol. 51, no. 5, Aug., pp. 331–334, 1983.

47. Tonomura, A., Osakabe, N., Matsuda, T., Kawasaki, T., Endo, J., Yano, S. & Yamada, H., "Evidence for Aharonov–Bohm effect with magnetic field completely shielded from electron wave," *Phys. Rev. Lett.*, Vol. 56, no. 8, Feb., pp. 792–795, 1986.

48. Tonomura, A. & Callen, E., "Phase, electron holography, and a conclusive demonstration of the Aharonov–Bohm effect," *ONRFE Sci. Bul.*, Vol. 12 no. 3, pp. 93–104, 1987.

49. Berry, M.V., "Exact Aharonov–Bohm wavefunction obtained by applying Dirac's magnetic phase factor," *Eur. J. Phys.*, Vol. 1, pp. 240–244, 1980.

50. Peshkin, M., "The Aharonov–Bohm effect: Why it cannot be eliminated from quantum mechanics," *Phys. Rep.*, Vol. 80, pp. 375–386, 1981.

51. Olariu, S. & Popescu, I.I., "The quantum effects of electromagnetic fluxes," *Rev. Mod. Phys.*, Vol. 157, no. 2, Apr., pp. 349–436, 1985.

52. Horvathy, P.A., "The Wu-Yang factor and the non-Abelian Aharonov–Bohm experiment," *Phys. Rev.*, Vol. D33, no. 2, 15 Jan., pp. 407–414, 1986.

53. Peshkin, M. & Tonomura, A. *The Aharonov–Bohm Effect* (Springer-Verlag, New York, 1989).

54. Aharonov, Y. & Casher, A., "Topological quantum effects for neutral particles," *Phys. Rev. Lett.*, Vol. 53, no. 4, 23 Jul., pp. 319–321, 1984.

55. Botelho, L.C.L. & de Mello, J.C., "A non-Abelian Aharonov–Bohm effect in the framework of pseudoclassical mechanics," *J. Phys. A: Math. Gen.*, Vol. 18, pp. 2633–2634, 1985.

56. Schulman, L.S., *Techniques and Applications of Path Integration* (Wiley, New York, 1981).

57. Bernido, C.S. & Inomata, A., "Path integrals with a periodic constraint: The Aharonov–Bohm effect," *J. Math. Phys.*, Vol. 22, no. 4, Apr., pp. 715–718, 1981.

58. Aharonov, Y., "Non-local phenomena and the Aharonov–Bohm effect," *Proc. Int. Symp. Foundation of Quantum Mechanics*, Tokyo, 1983, 10–19.

59. Sundrum, R. & Tassie, L.J., "Non-Abelian Aharonov–Bohm effects, Feynman paths, and topology," *J. Math. Phys.*, Vol. 27, no. 6, Jun., pp. 1566–1570, 1986.

60. Webb, R.A., Washburn, S., Umbach, C.P. & Laibowitz, R.B., "Observation of h/e Aharonov–Bohm oscillations in normal-metal rings," *Phys. Rev. Lett.*, Vol. 54, no. 25, 24 Jun., pp. 2696–2699, 1985.

61. Webb, R. A., Washburn, S., Benoit, A.D., Umbach, C.P. & Laibowitz, R.B., "The Aharonov–Bohm effect and long-range phase coherence in normal metal rings," *Proc. 2nd Int. Symp. Foundations of Quantum Mechanics 1986*, M. Namiki *et al.* (eds.) (The Physical Society of Japan, Tokyo), 1987, pp. 193–206.

62. Benoit, A.D. Washburn, S., Umbach, C.P., Laibowitz, R.B. & Webb, R.A., "Asymmetry in the magnetoconductance of metal wires and

loops," *Phys. Rev. Lett.*, Vol. 57, no. 14, Oct., pp. 1765–1768, 1986.

63. Washburn, S., Umbach, C.P., Laibowitz, R.B. & Webb, R.A., "Temperature dependence of the normal-metal Aharonov–Bohm effect," *Phys. Rev.*, Vol. B32, no. 7, Oct, pp. 4789–4792, 1985.

64. Washburn, S., Schmid, H., Kern, D. & Webb, R.A., "Normal-metal Aharonov–Bohm effect in the presence of a transverse electric field," *Phys. Rev. Lett.*, Vol. 59, no. 16, Oct, pp. 1791–1794, 1987.

65. Chandrasekhar, V., Rooks, M.J., Wind, S. & Prober, D.E., "Observation of Aharonov–Bohm electron interference effects with periods h/e and h/2e in individual micron-size, normal-metal rings," *Phys. Rev. Lett.*, Vol. 55, no. 15, 7 Oct., pp. 1610–1613, 1985.

66. Datta, S., Melloch, M.R., Bandyopadhyay, S., Noren, R., Vaziri, M., Miller, M. & Reifenberger, R., "Novel interference effects between parallel quantum wells," *Phys. Rev. Lett.*, Vol. 55, no. 21, 18 Nov., pp. 2344–2347, 1985.

67. Cavalloni, C. & Joss, W., "Aharonov–Bohm effect and weak localization in cylindrical Ag films," *Phys. Rev.*, Vol. B35, no. 14, 15 May, pp. 7338–7349, 1987.

68. Sandesara, N.B. & Stark, R.W., "Macroscopic quantum coherence and localization for normal-state electrons in Mg," *Phys. Rev. Lett.*, Vol. 53, no. 17, pp. 1681–1683, 1984.

69. Datta, S., Melloch, M.R., Bandyopadhyay, S. & Lundstrom, M.S., "Proposed structure for large quantum interference effects," *Appl. Phys. Lett.*, Vol. 48, no. 7, 17 February, pp. 487–489, 1986.

70. Datta, S. & Bandyopadhyay, S., "Aharonov–Bohm effect in semiconductor microstructures," *Phys. Rev. Lett.*, Vol. 58, no 7, 16 Feb., pp. 717–720, 1987.

71. Timp, G., Chang, A.M., Cunningham, J.E., Chang, T.Y., Mankiewich, P., Behringer, R. & Howard, R.E., "Observation of the Aharonov–Bohm effect for $\omega ct > 1$," *Phys. Rev. Lett.*, Vol. 58, no. 26, 29 Jun., pp. 2814–2817, 1987.

72. Sharvin, D.Y. & Sharvin, Y.V., "Magnetic flux quantization in a cylindrical film of a normal metal," *JETP Lett.*, Vol. 34, pp. 272–275, 1981.

73. Gijs, M., Van Haesendonck, C. & Bruynseraede, Y., Resistance oscillations and electron localization in cylindrical Mg films," *Phys. Rev. Lett.*, Vol. 52, no. 23, 4 Jun., pp. 2069–2072, 1984.

74. Altshuler, B.L., Aronov, A.G., Spivak, B.Z., Sharvin, Y.D. & Sharvin, Y.V., "Observation of the Aharonov–Bohm effect in hollow metal cylinders," *JETP Lett.*, Vol. 35, pp. 588–591, 1982.

75. Pannetier, B., Chaussy, J., Rammal, R. & Gandit, P., "Magnetic flux quantization in the weak-localization regime of a nonsuperconducting metal," *Phys. Rev. Lett.*, Vol. 53, no. 7, 13 Aug., pp. 718–721, 1984.

76. Bishop, D.J., Licini, J.C. & Dolan, G.J., "Lithium quenched-condensed microstructures and the Aharonov–Bohm effect," *Appl. Phys. Lett.*, Vol. 46, no. 10, pp.1000–1002, 1985.

77. Umbach, C.P., Van Haesendonck, C., Laibowitz, R.B., Washburn, S. & Webb, R.A., "Direct observations of ensemble averaging of the Aharonov–Bohm effect in normal metal loops," *Phys. Rev. Lett.*, Vol. 56, no. 4, 27 Jan., pp. 386–389, 1986.

78. Bandyopadhyay, S., Datta, S. & Melloch, R., "Aharonov–Bohm effect in semiconductor microstructures: Novel device possibilities," *Superlattices and Microstructures*, Vol. 2, no. 6, pp. 539–542, 1986.

79. Milliken, F.P., Washburn, S., Umbach, C.P., Laibowitz, R.B. & Webb, R.A., "Effect of partial phase coherence on Aharonov–Bohm oscillations in metal loops," *Phys. Rev.*, Vol. B36, no. 8, 15 Sep., pp. 4465–4468, 1987.

80. Edwards, J.T. & Thouless, D.J., "Numerical studies of localization in disordered systems," *J. Phys.*, Vol. C5, pp. 807–820, 1972.

81. Lee, P.A., Stone, A.D. & Fukuyama, H. "Universal conductance fluctuations in metals: Effects of finite temperature interactions and magnetic fields," *Phys. Rev.*, Vol. B35, no. 3, 15 Jan., pp. 1039–1070, 1987.

82. Stone, A.D. & Imry, Y., "Periodicity of the Aharonov–Bohm effect in normal-metal rings," *Phys. Rev. Lett.*, Vol. 56, no. 2, 13 Jan., pp. 189–192, 1986.

83. Imry, Y., "Active transmission channels and universal conductance fluctuations," *Europhys. Lett.*, Vol. 1, no. 5, 1 Mar., pp. 249–256, 1986.

84. Landauer, R., "Electrical resistance of disordered one-dimensional lattices," *Phil. Mag.*, Vol. 21, pp. 863–867, 1970.

85. Timp, G., Mankiewich, P.M., deVegvar, Behringer, R., Cunningham, J.E., Howard, R.E., Baranger, H.U. & Jain, J.K., "Suppression of the Aharonov–Bohm effect in the quantized Hall regime," *Phys. Rev.*, Vol. B39, pp. 6227–6230, 1989.

86. Dupuis, N. & Montambaux, G., "Aharonov–Bohm flux and statistics of energy levels in metals," *Phys. Rev. B*, Jun., 1991, in press.

87. Altshuler, B.L., Aronov, A.G. & Spivak, B.Z., "The Aharonov–Bohm effect in disordered conductors," *JETP Lett.*, Vol. 33, pp. 94–97, 1981.

88. Büttiker, M., Imry, Y. & Landauer, R., "Josephson behavior in small normal one-dimensional rings," *Phys. Lett.*, Vol. 96A, no. 7, 18 Jul., pp. 365–368, 1983.

89. Büttiker, M., Imry, Y., Landauer, R. & Pinhas, S., "Generalized many-channel conductance formula with application to small rings," *Phys. Rev.*, Vol. B31, no. 10, 15 May, pp. 6207–6215, 1985.

90. Xie, X.C. & DasSarma, S., "Aharonov–Bohm effect in the hopping conductivity of a small ring," *Phys. Rev.*, Vol. B36, 15 Dec., pp. 9326–9328, 1987.

91. Washburn, S. & Webb, R.A., "Aharonov–Bohm effect in normal metal: Quantum coherence and transport," *Adv. Phys.*, Vol. 35, no. 4, Jul.-Aug., pp. 375–422, 1986.

92. Polyarkov, Y.B., Kontarev, V.Y., Krylov, I.P. & Sharvin, Y.V., "Observation of quantum oscillations of the magnetoresistance of multiply connected objects with a hopping conductivity," *JETP Lett.*, Vol. 44, pp. 373–376, 1986.

93. Stone, A.D., "Magnetoresistance fluctuations in mesoscopic wires and rings," *Phys. Rev. Lett.*, Vol. 54, no. 25, 24 Jun., pp. 2692–2695, 1985.

94. Beenakker, C.W.J., "Theory of Coulomb-blockade oscillations in the conductance of a quantum dot," *Phys. Rev. B*, Jul., 1991, in press.

95. Beenakker, C.W.J., van Houton, H. & Staring, A.A.M., "Influence of Coulomb repulsion on the Aharonov–Bohm effect in a quantum dot," *Phys. Rev. B*, Jul., 1991, in press.

96. Aronov, A.G. & Sharvin, Y.V., "Magnetic flux effects in disordered conductors," *Rev. Mod. Phys.*, Vol. 59, pp. 755–779, 1987.

97. Boulware, D.G. & Deser, S., "The Aharonov–Bohm effect and the mass of the photon," manuscript, 1989.

98. Wu, T.T. & Yang, C.N., "Concept of nonintegrable phase factors and global formulation of gauge fields," *Phys. Rev.*, Vol. D12, no. 12, 15 Dec., 3843–3857, 1975.

99. Berry, M.V., "Quantal phase factors accompanying adiabatic changes," *Proc. R. Soc. Lond.*, Vol. A392, pp. 45–57, 1984.

100. Berry, M.V., "The adiabatic limit and the semiclassical limit," *J. Phys. A: Math. Gen.*, Vol. 17, pp. 1225–1233, 1984.

101. Berry, M.V., "Classical adiabatic angles and quantal adiabatic phase," *J. Phys. A: Math. Gen..* Vol. 18, pp. 15–27, 1985.

102. Berry, M.V., "Interpreting the anholonomy of coiled light," *Nature*, Vol. 326, 19 Mar., pp. 277–278, 1987.
103. Berry, M.V., "Quantum phase corrections from adiabatic iteration," *Proc. Roy. Soc. Lond.*, Vol. A414, pp. 31–46, 1987.
104. Wilkinson, M., "An example of phase holonomy in WKB theory," *J. Phys. A: Math. Gen.*, Vol. 17, pp. 3459–3476, 1984.
105. Wilkinson, M., "Critical properties of electron eigenstates in incommensurate systems," *Proc. R. Soc. Lond.*, Vol. A391, pp. 305–350, 1984.
106. Chiao, R.Y. & Wu, Y-S. "Manifestations of Berry's topological phase for the photon," *Phys. Rev. Lett.*, Vol. 57, no. 8, Aug., pp. 933–936, 1986.
107. Haldane, F.D.M., "Comment on 'Observation of Berry's topological phase by use of an optical fiber,'" *Phys. Rev. Lett.*, Vol. 59 no. 15, 12 Oct., pp. 1788–1790, 1987.
108. Chiao, R.Y. & Tomita, A., "Reply to Comments," *Phys. Rev. Lett.*, Vol. 59, no. 15, 12 Oct., pp. 1789, 1987.
109. Tomita, A. & Chiao, R.Y., "Observation of Berry's topological phase by use of an optical fiber," *Phys. Rev. Lett.*, Vol. 57, no. 8, Aug, pp. 937–940, 1986.
110. Simon, B., "Holonomy, the quantum adiabatic theorem, and Berry's phase," *Phys. Rev. Lett.*, Vol. 51 no. 24, Dec., pp. 2167–2170, 1983.
111. Thomas, T.F., "Direct calculation of the Berry phase for spins and helicities," manuscript, 1988.
112. Cai, Y.Q., Papini, G. & Wood, W.R., "On Berry's phase for photons and topology in Maxwell's theory," *J. Math. Phys.*, Vol. 31, pp. 1942–1946, 1990.
113. Aharonov, Y. & Anandan, J., "Phase change during a cyclic quantum evolution," *Phys.Rev. Lett.*, Vol. 58, no. 16, 20 Apr., pp. 1593–1596, 1987.
114. Bhandari, R. & Samuel, J., "Observation of topological phase by use of a laser interferometer," *Phys. Rev. Lett.*, Vol. 60, no. 13, 28 Mar., pp. 1211–1213, 1988.
115. Pancharatnam, S., *Proc. Indian Acad. Sci.*, Vol. A44, p. 247, 1956, reprinted in *Collected Works of S. Pancharatnam*, edited by G.W. Series (Oxford University Press, New York, 1975).
116. Kiritis, E., "A topological investigation of the quantum adiabatic phase," *Commun. Math. Phys.*, Vol. 111, pp. 417–437, 1987.

117. Jiao, H., Wilkinson, S.R., Chiao, R.Y. & Nathel, H., "Two topological phases in optics by means of a nonplanar Mach–Zehnder interferometer. *Phys. Rev.*, Vol. A39 pp. 3475–3486, 1989.

118. Suter, D., Chingas, G.C., Harris, R.A. and Pines, A., "Berry's phase in magnetic resonance," *Mol. Phys.*, Vol. 61, no. 6, pp. 1327–1340, 1987.

119. Suter, D., Mueller, K.T. and Pines, A., "Study of the Aharonov–Bohm quantum phase by NMR interferometry," *Phys. Rev. Lett.*, Vol. 60, no. 13, 28 Mar., pp. 1218–1220, 1988.

120. Richardson, D.J., Kilvington, A.I., Green, K. & Lamoreaux, S.K., "Demonstration of Berry's phase using stored ultracold neutrons," *Phys. Rev. Lett.*, Vol. 61, pp. 2030–2033, 1988.

121. Giavarini, G., Gozzi, E., Rohrlich, D. & Tacker, W.D., "On the removability of Berry's phase," *Phys. Rev.*, Vol. D39, pp. 3007–3015, 1989.

122. Giavarini, G. & Onofri, E., "Generalized coherent states and Berry's phase," *J. Math. Phys.*, Vol. 30, pp. 659–663, 1989.

123. Ellinas, D., Barnett, S.M. & Dupertius, M.A., "Berry's phase in optical resonance," *Phys. Rev.*, Vol. A39, pp. 3228–3237, 1989.

124. Gerry, C.C., "Berry's phase in the degenerate parametric amplifier," *Phys. Rev.*, Vol. A39, pp. 3204–3207, 1989.

125. Bialynicki-Birula, I. & Bialynicka-Birula, Z., "Berry's phase in the relativistic theory of spinning particles," *Phys. Rev.*, Vol. D35, pp. 2383–2387, 1987.

126. Kitano, M., Yabuzaki, T. & Ogawa, T., "Comment on 'Observations of Berry's topological phase by use of an optical fiber,'" *Phys. Rev. Lett.*. Vol. 58, no. 5, Feb., p. 523, 1987.

127. Chow, W.W., Gea-Banacloche, J., Pedrotti, L.M., Sanders, V.E., Schleich, W. & Scully, M.O., "The ring laser gyro," *Rev. Mod. Phys.*, Vol. 57, no. 1, Jan., pp. 61–104, 1985.

128. Hannay, J.H., "Angle variable holonomy in adiabatic excursion of an integrable Hamiltonian," *J. Phys. A: Math. Gen.*, Vol. 18, pp. 221–230, 1985.

129. Chiao, R.Y., Antaramian, A., Ganga, K.M., Jiao, H., Wilkinson, S.R., & Nathel, "Observation of a topological phase by means of a nonplanar Mach–Zehnder interferometer," *Phys. Rev. Lett.*, Vol. 60, no. 13, 28 Mar., pp. 1214–1217, 1988.

130. Segert, J., "Photon Berry's phase as a classical topological effect," *Phys. Rev.*, Vol. 36, no. 1, Jul. 1, pp. 10–15, 1987.

131. Segert, J., "Non-Abelian Berry's phase, accidental degeneracy, and angular momentum," *J. Math. Phys.*, Vol. 28, no. 9, Sep., pp. 2102–2114, 1987.

132. Bohren, C.F., "Scattering of electromagnetic waves by an optically active spherical shell," *J. Chem. Phys.*, Vol. 62, pp. 1566–1570, 1975.

133. Penrose, R. & Rindler, W., *Spinors and Space–Time*, 2 volumes (Cambridge University Press, 1984 & 1986).

134. Bialynicki-Birula, I. & Bialynicka-Birula, Z., *Quantum Electrodynamics* (Pergamon, New York, 1975).

135. Anandan, J. & Stodolsky, L., "Some geometrical considerations of Berry's phase," *Phys. Rev.*, Vol. D35, pp. 2597–2599, 1987.

136. Lévay, P., "Geometrical description of SU(2) Berry phases," *Phys. Rev.*, Vol. A41, pp. 2837–2840, 1990.

137. Layton, E., Huang, Y. & Chu, S-I., "Cyclic quantum evolution and Aharonov–Anandan geometric phases in SU(2) spin-coherent states, *Phys. Rev.*, Vol. 41, pp. 42–48, 1990.

138. Wang, S-J., "Nonadiabatic Berry's phase for a spin particle in a rotating magnetic field," *Phys. Rev.* Vol. A42, pp. 5107–5110, 1990.

139. Delacrétaz, G. Grant, E.R., Whetten, R.L., Wöste, L. & Zwanziger, J.W., "Fractional quantization of molecular pseudorotation in Na_3," *Phys. Rev. Lett.*, Vol. 56, no. 24, 16 Jun., pp. 2598–2601, 1986.

140. Herzberg, G. & Longuet-Higgens, H.C., "Intersection of potential energy surfaces in polyatomic molecules," *Disc. Faraday Soc.*, Vol. 35, pp. 77–82, 1963.

141. Longuet-Higgens, H.C., "The intersection of potential energy surfaces in polyatomic molecules," *Proc. Roy. Soc. Lond. Ser. A*, Vol. 344, pp. 147–156, 1975.

142. Mead, C.A. & Truhlar, D.G., "On the determination of Born–Oppenheimer nuclear motion wave functions including complications due to conical intersections and identical nuclei," *J. Chem. Phys.*, Vol. 70 pp. 2284–2296, 1979.

143. Nikam, R.S. & Ring, P., "Manifestation of Berry phase in diabolic pair transfer in rotating nuclei," *Phys. Rev. Lett.*, Vol. 58, no. 10, 9 Mar., pp. 980–983, 1987.

144. Bitter, T. & Dubbers, D., "Manifestation of Berry's topological phase in neutron spin rotation," *Phys. Rev. Lett.*, Vol. 59, pp. 251–254, 1987.

145. Goddard, P. & Olive, D.I., "Magnetic monopoles in gauge field theories," *Rep. Prog. Phys.* Vol. 41, pp. 1357–1437, 1978.

146. Martinis, J.M., Devoret, M.H. & Clarke, J., "Experimental tests for the quantum behavior of a macroscopic degree of freedom: The phase difference across a Josephson junction," *Phys. Rev.*, Vol. B35, no. 10, Apr., pp. 4682–4698, 1987.

147. Clarke, J. Cleland, A.N., Devoret, M.H., Esteve, D. & Martinis, J.M., "Quantum mechanics of a macroscopic variable: The phase difference of a Josephson junction," *Science*, Vol. 239, pp. 992–997, 1988.

148. Ehrenberg, E. & Siday, R.E., "The refractive index in electron optics and the principles of dynamics," *Proc. Phys. Soc., Lond.*, Vol. B62, no. 8, pp. 8–21, 1949.

149. Josephson, B.D., "Possible new effects in superconductive tunneling," *Phys. Lett.*, Vol. 1, no. 7, 1 Jul., pp. 251–253, 1962.

150. Josephson, B.D., "Coupled superconductors," *Rev. Mod. Phys.*, Vol. 36, Jan., pp. 216–220, 1964.

151. Josephson, B.D., "Supercurrents through barriers," *Adv. Phys.*, Vol. 14, pp. 419–451, 1965.

152. Josephson, B.D., "The discovery of tunneling supercurrents," *Science*, 3 May, Vol. 184, no. 4136, pp. 527–530, 1974.

153. Bloch, F., "Simple interpretation of the Josephson effect," *Phys. Rev.*, Vol. 21, no. 17, Oct., pp.1241–1243, 1968.

154. Barrett, T.W., "Nonlinear physics of electronic and optical materials for submicron device applications," Chap. 5, pp. 24–48 in T.W. Barrett & H.A. Pohl (eds.) *Energy Transfer Dynamics* (Springer Verlag, New York, 1987).

155. Lebwohl, P. & Stephen, M.J., "Properties of vortex lines in superconducting barriers," *Phys. Rev.*, Vol. 163, no. 2, 10 Nov., pp. 376–379, 1967.

156. Jaklevic, R.C., Lambe, J., Mercereau, J.E. & Silver, A.H., "Macroscopic quantum interference in superconductors," *Phys. Rev.*, Vol. 140, no. 5A, Nov., pp. A1628–1637, 1965.

157. von Klitzing, K., Dorda, G. & Pepper, M., "New method for high-accuracy determination of the fine-structure constant based on quantized Hall resistance," *Phys. Rev. Lett.*, Vol. 45, no. 6, Aug., pp. 494–497, 1980.

158. Stormer, H.L. & Tsui, D.C., "The quantized Hall effect," *Science*, Vol. 220, Jun., pp. 1241–1246, 1983.

159. Tsui, D.C., Stormer, H.L. & Gossard, A.C., "Two-dimensional magnetotransport in extreme quantum limit," *Phys. Rev. Lett.*, Vol. 48, no. 22, May, pp. 1559–1562, 1982.

160. Laughlin, R.B., "Quantized Hall conductivity in two dimensions," *Phys. Rev.*, Vol. B23, no. 10, May, pp. 5632–5633, 1981.

161. Post, E.J., "Can microphysical structure be probed by period integrals?", *Phys. Rev.*, Vol. D25, no. 12, Jun., pp. 3223–3229, 1982.

162. Post, E.J., "Electroflux quantization," *Phys. Lett.*, Vol. 92, no. 5, Nov., pp. 224–226, 1982.

163. Post, E.J., "On the quantization of the Hall impedance," *Phys. Lett.*, Vol. 94A, no. 8, Mar., pp. 343–345, 1983.

164. Aoki, H. & Ando, T., "Universality of the quantum Hall effect: Topological invariant and observable," *Phys. Rev. Lett.*, Vol. 57, no. 24, 15 Dec., pp. 3093–3096, 1986.

165. Byers, N. & Yang, C.N., "Theoretical considerations concerning quantized magnetic flux in superconducting cylinders," *Phys. Rev. Lett.*, Vol. 7, no. 2, Jul., pp. 46–49, 1961.

166. Sagnac, G., "L'éther lumineux demonstré par l'effet du vent relatif d'éther dans un interféromètre en rotation uniforme," *Comptes Rendus*, Vol. 157, Séance du 27 Oct., pp. 708–710, 1913.

167. Sagnac, G., "Sur la preuve de la realite de l'ether lumineaux par l'experience de l'interferographe tournant," *Comptes. Rendus*, Vol. 157, Seance du 22 Dec., pp. 1410–1413, 1913.

168. Sagnac, G., "Circulation of the ether in rotating interferometer," *Le Journal de Physique et Le Radium*, 5th Ser., Vol. 4, pp. 177–195, 1914.

169. Silvertooth, E.W., "Experimental detection of the ether," *Speculations in Science & Technology*, Vol. 10, no. 1, pp. 3–7, 1986.

170. Forder, P.W., "Ring gyroscopes: An application of adiabatic invariance," *J. Phys. A. Math. Gen.*, Vol. 17, pp. 1343–1355, 1984.

171. Michelson, A.A., "New measurement of the velocity of light," *Nature*, Vol. 114, Dec. 6, p. 831, 1924.

172. Michelson, A.A., "The effect of the earth's rotation on the velocity of light — Part I," *Astrophysical J.*, Vol. 61, no. 3, Apr., pp. 137–139, 1925.

173. Michelson, A.A., Gale, H.G. & Pearson, F., "The effect of the earth's rotation on the velocity of light — Part II," *Astrophys. J.*, Vol. 61, no. 3, Apr., pp. 140–145, 1925.

174. Michelson, A.A. & Morley, E.W., "Influence of motion of the medium on the velocity of light," *Am. J. Sci.*, Vol. 31, Series 3, pp. 377–386, 1886.

175. Post, E.J., "Sagnac effect," *Rev. Mod. Phys.*, Vol. 39, no. 2, Apr., pp. 475–493, 1967.

176. Post, E.J., "The Gaussian version of Ampere's law," *J. Math. Phys.*, Vol. 19, no. 1, Jan., p. 347, 1978.

177. van Dantzig, D., "The fundamental equations of electromagnetism, independent of metrical geometry," *Proc. Camb. Philos. Soc.*, Vol. 30, pp. 421–427, 1934.

178. Post, E.J., "Ramifications of flux quantization," *Phys. Rev.*, Vol. D9, no. 9, 15 Jun., pp. 3379–3386, 1974.

179. Post, E.J., *Formal Structure of Electromagnetics* (North Holland, Amsterdam, 1962).

180. Post, E.J., "The constitutive map and some of its ramifications," *Anns. Phys.*, Vol. 71, pp. 497–518, 1972.

181. Post, E.J., "Interferometric path-length changes due to motion," *J. Opt. Soc. Am.*, Vol. 62, pp. 234–239, 1972.

182. Born, M., "Reciprocity theory of elementary particles," *Rev. Mod. Phys.*, Vol. 21, no. 3, Jul., pp. 463–473, 1949.

183. Ali, S.T., "Stochastic localization, quantum mechanics on phase space and quantum space-time," *La Rivista del Nuovo Cimento*, Vol. 8, no. 11, pp. 1–128, 1985.

184. Ali, S.T. & Prugovecki, E., "Mathematical problems of stochastic quantum mechanics: Harmonic analysis on phase space and quantum geometry," *Acta Applicandae Mathematicae*, Vol. 6, pp. 1–18, 1986.

185. Hayden, H.C., "On a recent misinterpretation of Sagnac experiment," *Galilean Electrodynamics*, Vol. 2, pp. 57–58, 1991.

186. Hayden, H.C., "Yes, moving clocks run slowly, but is time dilated?", *Galilean Electrodynamics*, Vol. 2, pp. 63–66, 1991.

187. Yang, C.N., "Charge quantization, compactness of the gauge group, and flux quantization," *Phys. Rev.*, Vol. D1, no. 8, 15 Apr., p. 2360, 1970.

188. Yang, C.N., "Integral formalism for gauge fields, " *Phys. Rev. Lett.*, Vol. 33, no. 7, 12 Aug., pp. 445–447, 1974.

189. Dirac, P.A.M., "Quantized singularities in the electromagnetic field," *Proc. Roy. Soc. Lond.*, Vol. A133, pp. 60–72, 1931.

190. Dirac, P.A.M., "Theory of magnetic monopoles," *Phys. Rev.*, Vol. 74, pp. 817–830, 1948.

191. Weinberg, E.J., "Fundamental monopoles and multimonopole solutions for arbitrary simple gauge fields," *Nucl. Phys.*, Vol. B167, pp. 500–524, 1980.

192. Bogomol'nyi, E.B., "The stability of classical solutions," *Sov. J. Nucl. Phys.*, Vol. 24, no. 4, Oct., pp. 449–454, 1976.

193. Montonen, C. & Olive, D., "Magnetic monopoles as gauge particles," *Phys. Lett.*, Vol. 72B, no. 1, 5 Dec., pp. 117–120, 1977.

194. Atiyah, M. & Hitchin, N., *The geometry and Dynamics of Magnetic Monopoles* (Princeton University Press, 1980).

195. Craigie, N. (ed.), *Theory and Detection of Magnetic Monopoles in Gauge Theories (a Collection of Lecture Notes)* (World Scientific, Singapore, 1986).

196. Goldhaber, A.S. & Tower, W.P., "Resource letter MM-1: Magnetic monopoles," *Am. J. Phys.*, Vol. 58, pp. 429–439, 1990.

197. Schwinger, J., "Magnetic charge and quantum field theory," *Phys. Rev.*, Vol. 144, no. 4, 29 Apr., pp. 1087–1093, 1966.

198. Schwinger, J., "A magnetic model of matter," *Science*, Vol. 165, no. 3895, 22 Aug., pp. 757–761, 1969.

199. Wu, T.T. & Yang, C.N., "Dirac monopole without strings: Monopole harmonics," *Nucl. Phys.*, Vol. B107, pp. 365–380, 1976.

200. Wu, T.T. & Yang, C.N., "Dirac's monopole without strings: Classical Lagrangian theory," *Phys. Rev.*, Vol. D14, no. 2, 15 Jul., pp. 437–445, 1976.

201. 't Hooft, G., "Renormalizable Lagrangians for massive Yang–Mills fields," *Nuc. Phys.*, Vol. B35, pp. 167–188, 1971.

202. 't Hooft, G., "Magnetic monopoles in unified gauge theories," *Nucl. Phys.*, Vol. B79, pp. 276–284, 1974.

203. Polyakov, A.M., "Particle spectrum in quantum field theory," *JETP Lett.*, Vol. 20, no. 6, Sep. 20, pp. 194–195, 1974.

204. Prasad, M.K. & Sommerfeld, C.M., "Exact solution for the t'Hooft monopole and the Julia–Zee dyon," *Phys. Rev. Lett.*, Vol. 35, no. 12, 22 Sep., pp. 760–762, 1975.

205. Zeleny, W.B., "Symmetry in electrodynamics: A classical approach to magnetic monopoles," *Am. J. Phys.*, Vol. 59, pp. 412–415, 1991.

206. Higgs, P.W., "Broken symmetries, massless particles and gauge fields," *Phys. Lett.*, Vol. 12, no. 2, 15 Sep., pp. 132–133, 1964.

207. Higgs, P.W., "Broken symmetries and the masses of gauge bosons," *Phys. Rev. Lett.*, Vol. 13 no. 16, 19 Oct., pp. 508–509, 1964.

208. Higgs, P.W., "Spontaneous symmetry breakdown without massless bosons," *Phys. Rev.*, Vol. 145, no. 4, 27 May, pp. 1156–1163, 1966.

209. Brandt, R.A. & Primack, J.R., "Dirac monopole theory with and without strings," *Phys. Rev.*, Vol. 15, no. 4, Feb., pp. 1175–1177, 1977.

210. Yang, C.N., "Gauge fields, electromagnetism and the Bohm–Aharonov effect," *Proc. Int. Symp. Foundations of Quantum Mechanics* (Tokyo, 1983), pp. 5–9.

211. Gates. S.J., "Gauge invariance in nature: A simple view," pp. 31–80, in: R.E. Mickens (ed.) *Mathematical Analysis of Physical Systems* (Van Nostrand, New York, 1986).

212. Harmuth, H.F., "Correction for Maxwell's equations for signals I," *IEEE Trans. Electromagn. Compat.*, Vol. EMC-28, no. 4, Nov., pp. 250–258,1986.

213. Harmuth, H.F., "Correction for Maxwell's equations for signals II," *IEEE Trans. Electromagn. Compat.*, Vol. EMC-28, no. 4, Nov., pp. 259–265, 1986.

214. Harmuth, H.F., "Propagation velocity of Maxwell's equations," *IEEE Trans. Electromagn. Compat.*, Vol. EMC-28, no. 4, Nov., pp. 267–271, 1986.

215. Harmuth, H.F., "Reply to Djordjvic & Sarkar," *IEEE Trans. Electromagn. Compat.*, Vol. EMC-29, no. 3, Aug., p. 256, 1987.

216. Harmuth, H.F., "Reply to Wait," *IEEE Trans. Electromagn. Compat.*, Vol. EMC–29, no. 3, Aug., p. 257, 1987.

217. Harmuth, H.F., Boles, R.N. & Hussain, M.G.M., "Reply to Kuester's comments on the use of a magnetic conductivity," *IEEE Trans. Electromagn. Compat.*, Vol. EMC-29, no. 4, pp. 318–320, 1987.

218. Neatrour, M.J., "Comments on 'Correction of Maxwell's equations for Signals I,' 'Correction for Maxwell's equations for signals II,' and 'Propagation velocity of electromagnetic signals,'" *IEEE Trans. Electromagn. Compat.*, Vol. EMC-29, no. 3, pp. 258–259, 1987.

219. Wait, J.R., "Comments on 'Correction of Maxwell's equations for signals I and II,'" *IEEE Trans. Electromagn. Compat.*, Vol. EMC-29, no. 3, Aug., pp. 256–257, 1987.

220. Kuester, E.F., "Comments on 'Correction of Maxwell equations for signals I,' 'Correction of Maxwell equations for signals II,' and 'Propagation velocity of electromagnetic signals,'" *IEEE Trans. Electromag. Compat.*, Vol. EMC-29, no. 2, pp. 187–190, 1987.

221. Djordjvic, A.R. & Sarkar, T., "Comments on 'Correction of Maxwell's equations for signals I,' 'Correction of Maxwell's equations for signals II' and 'Propagation velocity of electromagnetic signals,'" *IEEE Trans. Electromagn. Compat.*, Vol. EMC-29, no. 3, pp. 255–256, 1987.

222. Hussain, M.G.M., "Comments on the comments of A.R. Djordjvic and T. Sarkar on 'Correction of Maxwell's equations for signals I and

II' and 'Propagation velocity of electromagnetic signals,"' *IEEE Trans. Electromagn. Compat.*, Vol. EMC–29, no. 4, p. 317, 1987.

223. Gray, J.E. & Boules, R.N., "Comment on a letter by E.F. Kuester," *IEEE Trans. Electromagn. Compat.*, Vol. EMC-29, no. 4, p. 317, 1987.

224. Barrett, T.W., "Comments on the Harmuth Ansatz: Use of a magnetic current density in the calculation of the propagation velocity of signals by amended Maxwell theory," *IEEE Trans. Electromagn. Compat.*, Vol. 30, pp. 419–420, 1988.

225. Barrett, T.W., "Comments on 'Solutions of Maxwell's equations for general nonperiodic waves in lossy media,"' *IEEE Trans. Electromagn. Compat.*, Vol. 31, pp. 197–199, 1989.

226. Barrett, T.W., "Comments on 'Some comments on Harmuth and his critics,"' *IEEE Trans. Electromagn. Compat.*, Vol. 31, pp. 201–202, 1989.

227. Hestenes, D., *New Foundations for Classical Mechanics* (Reidel, Dordrecht, 1987).

228. Hestenes, D. & Sobczyk, G., *Clifford Algebra to Geometric Calculus* (Reidel, Dordrecht, 1984).

229. Han, D. & Kim, Y.S., "Special relativity and interferometers," *Phys. Rev.*, Vol. A37, pp. 4494–4496, 1988.

230. Barrett, T.W., "Maxwell's theory extended. Part 1. Empirical reasons for questioning the completeness of Maxwell's theory, effects demonstrating the physical significance of the A potentials," *Annales de la Fondation Louis de Broglie*, Vol. 15, pp. 143–183, 1990.

231. Barrett, T.W., "Maxwell's theory extended. Part 2: Theoretical and pragmatic reasons for questioning the completeness of Maxwell's theory," *Annales de la Fondation Louis de Broglie*, Vol. 15, pp. 253–283, 1990.

232. Barrett, T.W., "The transition point between classical and quantum mechanical systems deterministically defined," *Physica Scripta*, Vol. 35, pp. 417–422, 1987.

233. Barrett, T.W., "Tesla's nonlinear oscillator-shuttle-circuit (OSC) theory," *Annales de la Fondation Louis de Broglie*, Vol. 16, pp. 23–41, 1991.

234. Barrett, T.W., *Energy Transfer Dynamics: Signal SU(2) Symmetry Conditioning for Radiation Propagation, Discrimination and Ranging* (W.J. Schafer Associates, June, 1987).

235. Yurke, B., McCall, S. L. & Klauder, J.R., "SU(2) and SU(1,1) interferometers," *Phys. Rev.*, Vol. A33, no. 6, Jun., pp. 4043–4054, 1986.

236. Paranjape, M.B., "Induced angular momentum and gauge invariance," *Phys. Rev.*, Vol. D36, no. 12, Dec., pp. 3766–3771, 1987.

237. Oh, C.H., Soo, C.P. & Lai, C.H., "The propagator in the generalized Aharonov–Bohm effect," *J. Math. Phys.*, Vol. 29, pp. 1154–1157, 1988.

238. Merzbacher, E., "Single valuedness of wave functions," *Am. J. Phys.*, Vol. 30, pp. 237–247, 1962.

239. Tellegen, B.D.H., "The gyrator: A new electric network element," *Phillips Research Reports*, Vol. 3, pp. 81, 1948.

240. Lakhtakia, A., "A case for magnetic sources," *Physics Essays*, Vol. 4, no. 1, pp. 105–108, 1991.

241. Lakhtakia, A., Varadan, A. & Varadan, V.V., *Time Harmonic Electromagnetic Fields in Chiral Media* (Springer Verlag, 1989).

242. Mikhailov, V.F., "Observation of the magnetic charge effect in experiments with ferromagnetic aerosols," *Annales de la Fondation Louis de Broglie*, Vol. 12, pp. 491–523, 1987.

243. Mikhailov, V.F., "On interpretation of the magnetic charge effect on ferromagnetic aerosols," *Academy of Sciences of Kazakh SSR*, Alma Ata, HEPI-88-05, 1988.

244. Mikhailov, V.F., "Observation of the Dirac magnetic charges by virtue of diffusion chamber," *Academy of Sciences of Kazakh SSR*, Alma Ata, HEPI-88-17, 1988.

245. Mikhailov, V.F., "Observation of apparent magnetic charges carried by ferromagnetic particles in water droplets," *J. Phys. A: Math. Gen.*, Vol. 24, pp. 53–57, 1991.

246. Mikhailov, V.F. & Mikhailova, L.I., "On some regularities of aerosol particle motion in electromagnetic fields," *J. Phys., A: Math. Gen.*, Vol. 23, pp. 53–63, 1990.

247. Dowker, J.S., "Quantum mechanics and field theory on multiply connected and on homogeneous spaces," *J. Phys., A. Gen. Phys.*, Vol. 5, Jul., pp. 936–943, 1972.

248. Schulman, L.S., "Approximate topologies," *J. Math. Phys.*, Vol. 12, no. 2, Feb., pp. 304–308, 1971.

249. Graneau, P., *Ampere–Neumann Electrodynamics of Metals* (Hadronic, Nonantum, Massachusetts, 1985).

250. Anderson, P.W., "Special effects in superconductivity" in *Lectures on the Many-Body Problem*, E.R. Caianello (ed.) (Academic, New York, 1964), Vol. 2, pp. 113–135.

251. Jackson, J.D., *Classical Electrodynamics*, 2nd edn. (Wiley, New York, 1975).
252. Panofsky, W.K.H. & Phillips, M., *Classical Electricity and Magnetism*, 2nd edn. (Addison-Wesley, Reading, Massachusetts, 1962).
253. Konopinski, E.J., "What the electromagnetic vector potential describes" *Am. J. Phys.*, Vol. 46, pp. 499–502, 1978.
254. Lenstra, D., Kamp, L.P.J. & Van Haeringen, W., "Mode conversion, Bloch oscillations and Zener tunneling with light by means of the Sagnac effect," *Optics Com.*, Vol. 60, pp. 339–344, 1986.
255. Wignall, J.W.G., "Frame dependence of the phase of de Broglie waves," *Am. J. Phys.*, Vol. 57, pp. 415–416, 1989.
256. Thirring, W., "A soluble relativistic field theory," *Ann. Phys.*, NY, Vol. 3, pp. 91–112, 1958.
257. Eisenhart, L.P., *Continuous Groups of Transformations* (Princeton University Press, 1933).
258. Hodge, W.V.D., *The Theory and Applications of Harmonic Integrals* (Cambridge University Press, 1959).
259. Palais, R.S., "The symmetries of solitons," *Bull. Am. Math. Soc.*, Vol. 34, 339–403, 1997.
260. Based on Barrett, T.W., "Electromagnetic phenomena not explained by Maxwell's equations," in A. Lakhtakia (ed.), *Essays on the Formal Aspects of Maxwell Theory* (World Scientific, 1993), pp. 8–86.

The Sagnac Effect: A Consequence of Conservation of Action Due to Gauge Field Global Conformal Invariance in a Multiply Joined Topology of Coherent Fields[i]

Overview

The Sagnac effect underlying the ring laser gyro is a coherent field effect and is described here as a global, not a local, effect in a multiply joined, not a simply joined, topology of those fields. Given a Yang–Mills or gauge field formulation of the electromagnetic field,[3] the measured quantity in the Sagnac effect is the phase factor. Gauge field formulation of electromagnetism requires in many cases uncoupling the electromagnetic field from the Lorentz group algebra. As conventionally interpreted, the Lorentz group is the defining algebraic

---

[i]Based on Ref. 4.

topology for the concepts of inertia and acceleration from an inertia-less state. However, those concepts find new definitions here in gauge theory and new group theory descriptions. The explanation for the origins of the Sagnac effect offered here lies in the generation of a *constraint* or *obstruction* in the interferometer's field topology under conditions of conserved action, that constraint or obstruction being generated *only* when the platform of the interferometer is rotated. Whereas previous explanations of the Sagnac effect have left the conventional Maxwell equations inviolate but seen a need to change the constitutive relations, here we see a need to do the opposite. Just as the Lorentz group description appears only as a limiting (*zero rotation or stationary*) case in this new explanation, so Minkowski space–time is also viewed as a limiting case appropriate for the Sagnac interferometer in the *stationary* platform situation, with Cartan–Weyl space–time appropriate for *rotated* platform situations. We attribute the existence of a measurable phase factor in the Sagnac interferometer with rotated platform to the conformal invariance of the action in the presence of the creation of a topological obstruction by the rotation.

1. Sagnac Effect Phenomenology

The Lorentz group algebra is the defining field algebra for the set of all *inertial* frames and the space–time symmetry. Any frame of reference that is not an inertial frame is an "accelerated" frame and is experienced as a force field. However, we shall attempt to show that for *noninertial* frames, the Lorentz group is *not* the defining algebra. Such situations are measured by rotation sensors. Of these, Sagnac[40–42] first demonstrated a ring interferometer which indicates the state of rotation of a frame of reference, i.e. a ring interferometer as a rotation rate sensor. The ring interferometer performs the same function as a mechanical gyroscope. When a laser is used as the source of radiation in the interferometer, it is called a ring laser gyro.

Figure 1.1 shows the basic Sagnac interferometer. One light beam circulates a loop in a clockwise direction, and another beam circulates a loop in a counterclockwise direction. When the interferometer is set in motion, interference fringes (phase difference) are observed at the overlap area H, i.e. in the heterodyned counterpropagating beams.

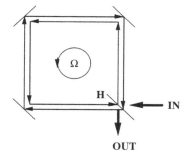

Fig. 1.1.   The basic Sagnac interferometer. The clockwise and counterclockwise beams are not separated in reality and are shown here only for purposes of exposition.

Details of the interferometer can be found in Refs. 8 and 30 and reviews of the Sagnac effect have been given by von Laue,[43] Zernicke[57] and Metz.[19–21] There are two basic kinds of ring interferometers that sense rotation: the passive ring resonator and the active ring laser gyro. The general theory of the ring laser gyro is addressed in Refs. 1, 11 and 18.

Here, we shall quickly cover the main descriptive features and move on to address the central issue of this paper: explanations of the effect. In the case of the passive ring resonator, the interference fringes are described by

$$\Delta\phi = \frac{4\Omega \cdot A}{\lambda c}, \qquad (1.1)$$

where $\Delta\phi$ is the phase difference between clockwise and counterclockwise propagating beams, Ω is the angular rate in rad/sec, λ is the vacuum wavelength, A is the area enclosed by the light path, and a velocity field v defines the angular velocity:

$$\nabla \times v = 2\Omega. \qquad (1.2)$$

Rosenthal[39] suggested a self-oscillating version of the Sagnac ring interferometer which was demonstrated by Macek and Davis.[17] In this version, the clockwise and counterclockwise modes occur in the same optical cavity. In the case of the laser version of the self-oscillating version of the Sagnac interferometer, i.e. the ring laser gyroscope, and in contrast to the Sagnac ring interferometer, a comoving optical medium in the laser beam affects the beat frequency, which, rather than the phase difference, is the measured variable. In this version, the frequency difference, $\Delta\omega$, of the clockwise and counterclockwise propagating beams with respect to the resonant frequency, ω, is described by

$$\left|\frac{\Delta\omega}{\omega}\right| = \frac{2 \oint n^2(1-\alpha)v \cdot dr}{c \oint n \, ds}, \tag{1.3}$$

where n is the index of refraction of the stationary medium, and α is a coefficient of drag. The same frequency difference with respect to the angular velocity is given by

$$\Delta\omega = \sqrt{\frac{A}{\pi}}\frac{2\Omega}{\lambda} = \frac{4A\Omega}{P\lambda}, \tag{1.4}$$

where P is the perimeter of the light path. The Sagnac interferometer path length change in terms of phase reversals is independent of the wave guide mode and completely independent of the optical properties of the path.[32]

There are details of ring laser gyroscope operation, such as lock-in and scale factor variation, which necessitate the amendment of the above descriptions, but these operational details will not be addressed here, and we move directly to consider explanations of the effect.

There are three current explanations/descriptions of the Sagnac effect:

(a) the kinematic description,
(b) the physical-optical description,
(c) the dielectric metaphorical description.

It is the intention of the present essay to introduce a fourth:

(d) the gauge field explanation.

We shall examine each of these approaches in turn.

1.1. *The kinematic description*

The force field exerted on the fields of the Sagnac interferometer can be either due to gravitational, linear acceleration or to rotational velocity (kinematic acceleration) field effects. The kinematic acceleration field is due to Coriolis force contributions, but the gravitational field and linear acceleration field do not have such contributions. There is also a distinction between the different fields with respect to the convergence/divergence of those fields. For example, the lines of gravitational force converge to a nonlocal point, e.g. the center of the earth, in the case of a platform at rest on the earth, but in the case of a linearly accelerated platform the lines of force converge to a nonlocal point at infinity. No Coriolis force is present in either of these cases. In contrast to these, with a platform with rotational velocity, the lines of force diverge from the local axis of rotation and a Coriolis force is present. If the platform with rotational velocity is also located on or near the earth, it will also experience the gravitational force of the earth, besides the Coriolis force. However, only the platform undergoing kinematic acceleration (rotational velocity) is in a state of motion with respect to *all* inertial frames.

The kinematic description thus primarily implies the Coriolis (acceleration) force and a state of kinematic acceleration is associated with a state of absolute motion with respect to all inertial frames. For example, Konopinski[16] defined the electromagnetic vector potential as field momentum exchanged with the kinetic momenta of charged particles. According to this author, most defining relations between potentials and fields, such as in equations of motion, are defined in a *static condition*. In these static conditions, and moreover *local conditions*, $q\phi$ can be defined as a "store" of field energy, and qA/c a "store" of momentum energy. That is the conventional interpretation of the four vector potential.

However, the subject of Konopinski's paper is a second condition which is *global* in character. The model he considers is not a *point* but a *volume* $\oint dV(r)$ of an electromagnetic field with an energy and vector momentum defined as in the Coulomb gauge,

$$w(r, t) = \frac{(E^2 + B^2)}{8\pi}, \tag{1.1.1}$$

$$g(r, t) = \frac{(E \times B)}{4\pi c}, \tag{1.1.2}$$

and with a total mass

$$W = \oint \frac{dVw}{c^2},$$

(1.1.3)

which is constant in the *absence* of fluxes through the surface enclosing the volume. The test condition for this situation consists of a point charge q at a fixed position r_q in a static external field $E_0 = -\nabla\phi$, $B_0 = -\nabla \times A(r)$. The fact that the model considered is a volume introduces *global* (across the volume) and *local* (within the volume, e.g. pointlike) conditions.

The equation of motion for a point charge is given as

$$\frac{d\left(Mv + \frac{q}{c}A\right)}{dt} = -\nabla q \left(\phi - \left(\frac{v}{c}\right) \cdot A\right),$$

(1.1.4)

which describes changes in conjugate momentum (left side) + interaction energy (right side). With q constant (the left hand side), any variation of v causes A to vary, and vice versa. The total field momentum that changes when the position r_q is changed is derived as

$$P(r_q) \equiv \oint dV(r) \frac{E_q(r - r_q) \times B_0(r)}{4\pi c}.$$

(1.1.5)

Introducing source terms gives the field momentum $P(r_q)$ as

$$P(r_q) = \frac{qA(r_q)}{c}.$$

(1.1.6)

The result is that under the total momentum conditions expressed by Eq. (1.1.4), changes in the total field momentum when the position of the particle at r_q is changed must result in changes in the kinetic momentum of the particle Mv.

Considering now the present topic of interest, the Sagnac effect, one may state the equilibrium condition in the reverse causal condition to that considered by Konopinski, namely, changes in the velocity of the system defining the kinetic energy of a "particle," v, result in changes in the vector potential in the total field momentum, A. Moreover, in the Sagnac effect there are *two* vector potential components with respect to clockwise and counterclockwise beams. The measured quantity, as will be explained more fully below, is then *the phase factor* or the integral of the *potential difference* between those beams and related to the *angular velocity difference* between the two beams. Therefore, as the vector potential measures the momentum gain and the scalar potential measures the kinetic energy gain, the photon will acquire "mass."[j]

Konopinski, using variational principles, formulated a Langrangian for this global field situation with A_ν as "generalized coordinates" and $\partial_\mu A_\nu$ as "generalized velocities."

$$L = -\frac{(\partial_\mu A_\nu)^2}{8\pi} + \frac{j_\nu A_\nu}{c}. \qquad (1.1.7)$$

With gauge invariance (the Lorentz gauge) there are no source terms, $j_\nu \equiv 0$, so

$$L = -\frac{(\partial_\mu A_\nu)^2}{8\pi}. \qquad (1.1.8)$$

Thus, in our adaptation of this argument, Eq. (1.1.8) describes the field conditions of the Sagnac interferometer *when its platform is stationary*, but conveys no more information than the field tensor $F_{\mu\nu} = \partial_\mu A_\nu - \partial_\nu A_\mu$. On the other hand, (i) the conservation condition expressed by Eqs. (1.1.4) and (1.1.7) describes the Sagnac interferometer platform *in rotation and kinetically*; and (ii) it is relevant that Eq. (1.1.8), but *not* Eq. (1.1.7), is determined by the Lorenz gauge.[k] Therefore the field algebraic logic underlying the Sagnac effect, i.e. the Sagnac interferometer platform *in rotation*, is *not* that of the Lorenz gauge.

1.2. *The physical–optical description*

Post's physical optical theory of the Sagnac effect[30,33] demands the loosening of the ties of the theory of electromagnetism to a Lorentz[l]-invariant structure. Post correctly observed that it is customary

---

[j]Moyer,[23] addressing the combining of electromagnetism and general relativity, identified charge with a Lagrange multiplier and the Hamiltonian with the self-energy mc^2, using an optimal control argument rather than the calculus of variations. The Lagrangian identified is a function of the electromagnetic scalar potential and the vector potential, i.e. the four-potential.

[k]Evidently, this gauge is due to L. Lorenz (1829–1891) of Copenhagen, not H.A. Lorentz (1853–1928) of Leiden (cf. Ref. 53 (1951), vol. 1, pp. 267–268, and Ref. 26, vol. 1, p. 321, footnote).

[l]H.A. Lorentz (1853–1928).

to assume no distinction in free space between dielectric displacement, *D*, and the electric field, *E*, or between magnetic induction, *B*, and the magnetic field, *H*. This identification is called the *Gaussian identification* and is justified by the supposed absence of polarization mechanisms in free space. The Gaussian identification, together with the Maxwell equations, leads to the d'Alembertian equation, which is a Lorentz invariant structure. As the d'Alembertian does not permit mixed space–time derivatives, it cannot account for the nonreciprocal asymmetry between clockwise and counterclockwise beam rotation in the Sagnac interferometer. Therefore, Post suggested that in order to account for this asymmetry, either the Gaussian field identification is incorrect in a rotating frame, or the Maxwell equations are affected by the rotation. However, in offering this choice, Post tacitly assumed no linkage between the Gaussian field identification and the Maxwell equations (i.e. their exclusivity was assumed). He also assumed that the solution to the asymmetry effect must be a *local* effect, because he was convinced that both the Gaussian field identification and the conventional Maxwell equations describe *local* effects. On the other hand, we argue below that the field arrangements in the interferometer should be described as a *global* situation and, as a result, the occurrence of an asymmetry does not warrant a change in the *local* Abelian Maxwell equations (also rejected by Post), but to the required use of *nonlocal*, non-Abelian Maxwell equations in a *multiply connected interferometric* situation (neglected by Post); and also does not warrant a change in the *local* Gaussian field identification (adopted by Post), but warrants the use of *nonlocal* non-Abelian field-metric interactions (neglected by Post). Both *nonlocal* non-Abelian equations and *nonlocal* interactions are required because the asymmetry under discussion arises in the Sagnac interferometer, this interferometer being a *global*, i.e. nonlocal, situation; and the amendment suggested by Post, whether of the local equations or the local identification, is inappropriately *ad hoc* in the presence of that *global* situation. Rather, the field topology of the Sagnac interferometer requires drastic redefinition of those equations and the full interaction logic, rather than a topologically inappropriate amendment of their local form regardless of field topology.

Nonetheless, Post understood that there was, and is, a problem in defining field-metric relations in empty space and referred to, among others, the Pegram[25] experiment as illustrative. Quickly stated, in this experiment, simultaneously and around the same axis, a coaxial cylindrical condenser and solenoid are rotated. The rotation produces a magnetic field in the solenoid in the axial direction and between the plates of the condensor. The condensor can then be charged by shortening the plates of the condenser. Post observed that the experiment indicates a cross-relation between electric and magnetic fields in a vacuum, a relation which is denied by a Lorentz transformation.

Post[30] also noticed correctly that Weyl and Cartan were aware of the metric independence of Maxwell's equations. But Post[34] presented the view that the asymmetry of the conventional Maxwell equations (absence of magnetic monopoles) is compatible with a certain topological symmetry, which he then used to suggest that the law of flux quantization, $\oint F = 0$, is a fundamental law. This suggestion, however, is contradicted by the law of flux quantization being *global*, rather than *local*, in nature, and being based on a *multiply connected* symmetry. Post's suggestion was motivated evidently by the assumption — incorrect from our perspective — that the space–time situation of the Sagnac interferometer is *simply connected*. He also distinguished two points of view:

(1) Topology enters physics through the families of integration manifolds that are generated by physical fields, and space–time is the arena in which these integration manifolds are embedded.
(2) Space–time is endowed with a topological structure relating to its physics.

The first point of view brings the *field* center stage; the second, the *metric*. However, from a gauge field perspective, these two points of view are neither unconnected nor exclusive and thus do not constitute an exclusive choice. Under a gauge field formulation, there *can be* interaction between field and metric. The exclusive choice offered by Post is understandable in that he did not distinguish between force fields and related gauge fields. From a gauge theory point of view, however, one is not forced to choose exclusively between these two alternatives.

Post also proposed that the global condition for the **A** potential to exist is that all cyclic integrals of F vanish. However, this statement is again based on consideration of *simply connected* domains, for which the local Maxwell theory is, indeed, described by $\oint F = 0$. He stated that flux quantization is formally incompatible with the magnetic monopole hypothesis, because of his belief that a global **A** exists only if $\oint F = 0$. But, as we shall show below, in a *multiply connected* domain in the presence of a topological obstruction, a *global phase factor* defined over local **A** fields exists. Therefore in this global, multiply connected situation, $\oint F \neq 0$.

Nonetheless, Post's physical optical theory was a major advance in understanding and is based on the following valid observations:

(1) The Maxwell equations have no specific constitutive relations to free space. The traditional equalities in free space, $\boldsymbol{E} = \boldsymbol{D}$ and $\boldsymbol{B} = \boldsymbol{H}$, assume the Gaussian approximation (absence of detected polarization mechanisms in free space) discussed above, and define the properties of free space *only as seen from inertial frames*. This is because those relations and the Maxwell equations lead to the standard free-space d'Alembertian wave equation, and the d'Alembertian, as Post pointed out, is a Lorentz-invariant structure. [The Gaussian approximation corresponds to the Minkowskian metric $(c^2, -1, -1, -1)$, or $(1, -1, -1, -1)$ with $(c^2 = 1)$, defining the Lorentz group as a symmetry property of the space-time continuum.]

(2) Only in the case of *uniformly translating* systems does the mutual motion of observer and platform *completely* define the physical situation.

Post's solution to the problems raised by these observations is, as we have seen, to modify the constitutive relations, but not to modify the field or Maxwell equations. This solution, we have suggested, inappropriately assumes that the Sagnac interferometer is a simply connected geometry. The Pegram experiment, described above, indicates the presence of cross-coupling between electric and magnetic fields. According to Post, this cross-coupling is responsible for the Sagnac effect and is due to the constitutive relations on a rotating frame. However, it is ironic that it is not possible to assume the physicality of such cross terms without modification of the Maxwell equations which Post left untouched. Therefore, while we may agree that the cross-coupling is related to the Sagnac effect, below we ascribe its existence to the presence of non-Abelian Maxwell relations,[3] rather than amended constitutive relations.

We can also agree that "A nonuniform motion produces a real and intrinsic physical change in the object in motion; the motion of the frame of reference by contrast produces solely a difference in the observational viewpoint." (Ref. 30, p. 488.) However, we would reword the distinction as follows: Whereas the *linear motion* of the frame of reference of an interferometer incorporating area *A* produces solely a *local* difference in the observational viewpoint describable by the Lorentz gauge, the *nonuniform motion* of an interferometer incorporating area *A* , as on a rotating platform, produces a *global* difference describable by the Ampère gauge.

Despite these — considered here incorrect — theoretical positions, Post[30] greatly advanced the understanding of the Sagnac effect by squarely addressing the physical issues. We shall return to these physical issues later. He was also a leader in realizing the necessity of distinguishing local versus global approaches to physics.[29,31,34,35].

1.3.   *The dielectric metaphor description*

Chow *et al.*[8] commenced their dielectric metaphor for the gravitational field with the Plebanski[28] observation that it is possible to write Maxwell's equations in an arbitrary gravitational field in a form in which they resemble electrodynamic equations in a dielectric medium. Thus, the gravitational field is in some sense equivalent to a dielectric medium and represented by the metric $g_{\mu\nu} = g_{\nu\mu}$ of the form

$$g_{\mu\nu} = \eta_{\mu\nu} + h_{\mu\nu} = \begin{bmatrix} 1 & 0 & 0 & 0 \\ 0 & -1 & 0 & 0 \\ 0 & 0 & -1 & 0 \\ 0 & 0 & 0 & -1 \end{bmatrix} + \begin{bmatrix} 0 & h_{01} & h_{02} & h_{03} \\ h_{10} & 0 & 0 & 0 \\ h_{20} & 0 & 0 & 0 \\ h_{30} & 0 & 0 & 0 \end{bmatrix},$$

$$(1.3.1)$$

where $\eta_{\mu\nu}$ is the metric of special relativity and $h_{\mu\nu}$ is the effect of the gravitational field. Using the definition

$$h \equiv (h_{01}, h_{02}, h_{03}), \tag{1.3.2}$$

Chow *et al.*[8] defined amended constitutive relations. That is, just like Post[30] examined above, these authors chose to introduce any metric influences on the fields into the constitutive relations rather than into amended Maxwell equations. The amended constitutive relations offered by these authors are

$$\boldsymbol{D} = \boldsymbol{E} - c(\boldsymbol{B} \times \boldsymbol{h}), \tag{1.3.3}$$

$$\boldsymbol{B} = \boldsymbol{H} + \frac{1}{c}(\boldsymbol{E} \times \boldsymbol{h}). \tag{1.3.4}$$

These authors then proceeded to derive an equation of motion for the electric field. However, as we have seen above, because an equation of motion (d'Alembertian) *assumes* a Lorenz gauge[m] (i.e. the special theory of relativity), and as $h_{\mu\nu}$ is introduced *as a correction to the special theory components*, $\eta_{\mu\nu}$, such a derivation must be at the cost of *group algebraic inconsistency*. Furthermore, if, as is claimed by many (commencing with Heaviside[12]) there is a formal analogy between the gravitational potential, $\boldsymbol{h}$, and the vector potential, $\boldsymbol{A}$, and the field $\nabla \times \boldsymbol{h}$ and the magnetic field $\boldsymbol{B}$, that analogy can nonetheless be introduced into the electromagnetic field in ways *other than* by the amended constitutive relations, (1.3.3) and (1.3.4). The next section addresses this other way.

1.4. *The gauge field explanation*

Having found the physical–optical and the dielectric metaphorical explanations of the Sagnac effect wanting, we now introduce a gauge theory[n] explanation, but before doing so we examine Forder's[10] analysis of ring gyroscopes. A presupposition of gauge theory is constant action and Forder[10] showed that the adiabatic invariance of the action

---

[m]L. Lorenz (1829–1891) of Copenhagen, not H.A. Lorentz (1853–1928) of Leiden.
[n]The concept of gauge was originally introduced by Weyl[46,48] to describe a local scale or metric invariance for a global theory (the general theory of relativity). This use was discontinued. Later, it was applied to describe a local phase invariance in quantum theory.

implies the invariance of the flux enclosed by the contour, which is Lenz's law. There is a common framework for treatment of (1) the ring laser gyro, (2) conductors and (3) superconductors. The treatment of (1) is according to the adiabatic invariance of the quantum phase:

$$\Delta\phi = 0 = L\Delta k + \omega\Delta T, \qquad (1.4.1)$$

where L is the length of the contour, and the treatments of (2) and (3) are according to the adiabatic invariance of the magnetic flux:

$$\Delta\Phi = 0 = L_0\Delta i + \frac{E}{q}\Delta T, \qquad (1.4.2)$$

where L_0 is the inductance of the contour.

Forder's thesis implies:

(1) treatment of the action as an adiabatic invariant when the gyro is subject to a slow angular acceleration;
(2) generalization of the action integral to a noninertial frame of reference which requires the general theory of relativity;
(3) defining particles on the contour (of a platform), which provide angular momentum and the action of their motion in the rotating frame, that angular momentum and action involving not only (a) the particles' (linear) momentum, but also (b) a term proportional to the particles' energy.

Forder's claim is that (b) distinguishes rotation-sensitive from rotation-insensitive devices. We quickly outline the main points of this claim.

With the intrinsic angular momentum defined,

$$I = \oint pdq \qquad (1.4.3)$$

(and the action = integral over one complete cycle), Forder's model is one of particles at the periphery of the frame at a distance l. If Hamilton's principal function is $S(t, l)$, the energy is E and the momentum is p for each particle, then

$$I = \oint \frac{\partial S}{\partial l}dl = \oint pdl, \qquad (1.4.4)$$

and the transit time around the contour is

$$T = \frac{dI}{dE}. \tag{1.4.5}$$

Free propagation is assumed for both particles and light propagating between perfectly reflecting surfaces arranged around the contour. Then

$$S(t, l) = pl - Et, \tag{1.4.6}$$

and

$$I = pL, \tag{1.4.7}$$

where $L = \oint dl$ is the length of the contour. The period of motion is then

$$T = \frac{dI}{dE} = L\frac{dp}{dE} = \frac{L}{v}, \tag{1.4.8}$$

where $v = \frac{dE}{dp}$ is the velocity of the particle.

If the contour and the observer are accelerated to an angular velocity Ω, the action changes to

$$I^A = \oint \frac{\partial S^A}{\partial x^i} dx^i = -\oint p_i dx^i \tag{1.4.9}$$

($i = 1, 2, 3$), where the p_i components are the spatial parts of the four-momentum

$$p_\mu = -\frac{\partial S^A}{\partial x^\mu}. \tag{1.4.10}$$

Forder's argument is that

$$I^A = I, \tag{1.4.11}$$

i.e. the action is an adiabatic invariant under the acceleration, and that

$$I^A = p^A L + E^A \Delta T, \tag{1.4.12}$$

where ΔT is a synchronization discrepancy between clocks at different points in a rotating frame. The action for a ring laser gyro is then simply

$$I^A = E^A T^A, \tag{1.4.13}$$

where

$$T^A = \oint \frac{dx^0}{c} = \frac{dI}{dE^A} = \frac{L dp^A}{dE^A} + \Delta T = \frac{L}{v^A} + \Delta T \tag{1.4.14}$$

is the transit time around the contour, and

$$v^A = \frac{dE^A}{dp^A} \tag{1.4.15}$$

is the proper velocity of the particle. Forder then claimed that although the proper velocity is c for all observers, in a rotating frame a photon takes a different length of time to traverse the contour. However, it is difficult to understand how this can be claimed. If the clockwise and counterclockwise beams in the Sagnac interferometer traverse *different* paths, then one might state that, under platform rotation, the physical distance changed to compensate for changes in ΔT, the time taken to traverse the length of the rotating interferometer back to point H (see Fig. 1.1). But this is not the case. It is the *same* interferometric path for both beams. So how it can shorten for one beam and lengthen for another is mysterious.

Leaving that aside: depending on the direction of propagation, the period is increased or decreased by ΔT, defined as

$$\Delta T = \frac{1}{c^2} \oint (\Omega \wedge r) \cdot dl = \frac{2}{c^2} \Omega \cdot S, \tag{1.4.16}$$

where S is the contour area, and a link was established by Forder between the Sagnac effect and action constancy — a constancy which underlies gauge theory, to which we now turn.

In the case of conventional electromagnetism, phase is arbitrary (there is gauge invariance) and fields (of force) are described completely by the electromagnetic field tensor, $f_{\mu\nu}$. This is the case when the theoretical model addresses only *local* effects. In the case of *parallel transport*, however, the phase of a wave function ψ representing a particle of charge e at point x is parallel to the phase at another point

$x + dx^\mu$ if the local values differ by

$$ea_\mu(x)dx^\mu, \tag{1.4.17}$$

where $\mu = 0, 1, 2, 3$ and $a_\mu(x)$ represents a set of functions. A *gauge transformation* at x with a phase change $e\alpha(x)$ is written as

at x:

$$\psi(x) \rightarrow \psi'(x) = \exp[ie\alpha(x)]\psi(x), \tag{1.4.18a}$$

but at $x + dx^\mu$ it is written as

at $x + dx^\mu$:

$$\psi(x + dx^\mu) \rightarrow \psi'(x + dx^\mu) = \exp[ie\{\alpha(x) + \partial_\mu\alpha(x)dx^\mu\}]\psi(x). \tag{1.4.18b}$$

So, in the case of a phase change *with phase parallelism*,[13]

$$a'_\mu(x) = a_\mu(x) + \partial_\mu(x)\alpha(x), \tag{1.4.19}$$

where ∂_μ signifies $\partial/\partial x^\mu$.

Figures 1.4.1–1.4.3 are examples of representations defining topologically parallel transport for two counterpropagating beams. Referring to Fig. 1.4.1, along either of the paths, a change in phase is given by

$$e \int_\Gamma^Q a_\mu(x)dx^\mu. \tag{1.4.20}$$

The difference between the phases at Q along two distinct paths is

$$e \int_{\Gamma_2}^Q a_\mu(x)dx^\mu - e \int_{\Gamma_1}^Q a_\mu(x)dx^\mu = e \oint_{\Gamma_2 - \Gamma_1} a_\mu(x)dx^\mu. \tag{1.4.21}$$

Use of Stokes' theorem obtains the surface integral

$$-e \iint f_{\mu\nu}(x)d\sigma^{\mu\nu} \tag{1.4.22}$$

over any surface Σ bounded by the closed curve $\Gamma_2 - \Gamma_1$ with

$$f_{\mu\nu} = \partial_\nu a_\mu(x) - \partial_\mu a_\nu(x). \tag{1.4.23}$$

If $f_{\mu\nu}$ is nonzero, then parallel transport of phases is path-dependent. But $f_{\mu\nu}$ is gauge-invariant (because as a difference, it is independent of any phase rotations at that point). Therefore $f_{\mu\nu}$ has the defining characteristics of the electromagnetic field tensor and the $a_i(x)$ of gauge

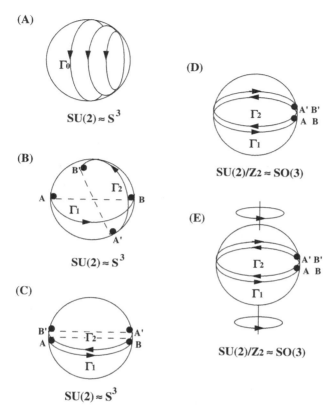

Fig. 1.4.1. Generation of a topological obstruction (source of magnetic flux) in the Sagnac interferometer. (A) Parallel transport along two or more different paths; (B) parallel transport along two different paths with a path reversal; (C) parallel transport along the same path; (D) parallel transport along the same path with a path reversal around an obstruction; (E) parallel transport along the same path with a path reversal around an obstruction and with twist. Paths taken by two counterpropagating beams are separated in C, D and E for purposes of exposition. The counterpropagating beams are superposed in reality with A = B′ and B = A′ in (C) and A = B = A′ = B′ in (D) and (E). SU(2) is a three-sphere S^3 in four-space; Z_2 are the integers modulo 2; SU(2)/Z_2 = SO(3) is a three-sphere in four-space with identity of pairs of opposite signs, e.g. $|\pm a| = \xi$. The twist in E corresponds to a "patch" condition, and in the Sagnac interferometer it is caused by the presence of angular velocity (+ linear acceleration), as shown in E.

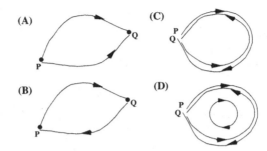

Fig. 1.4.2. Generation of a topological obstruction (source of magnetic flux) in the Sagnac interferometer. (A) two-beam parallel transport along two different paths from P to Q around an obstruction with progress in the same direction along the paths; (B) two-beam parallel transport along two different paths around an obstruction with a path reversal; (C) two-beam parallel transport along one path around an obstruction with P = −Q and Q' = −P; (D) two-beam parallel transport along one path around an obstruction with P = −Q and Q' = −P and with a twist. Paths taken by two counterpropagating beams are separated in C and D for purposes of exposition. The twist in D corresponds to a "patch" condition, and in the Sagnac interferometer it is caused by angular velocity (+ linear acceleration). The counterpropagating beams are superposed in reality. Figure 1.4.2(C) corresponds to Fig. 1.4.1(D); and Fig. 1.4.2(D) to Fig. 1.4.1(E).

potentials. This is as far as conventional Maxwell theory takes us, leaving the situation of a rotated platform shown in Figs. 1.4.1–1.4.3 underdescribed. To progress further requires Yang–Mills theory[55,56] to which we now turn.

Yang–Mills theory is a generalization of electromagnetism in which ψ, a wave function with two components, e.g. $\psi = \psi'(x)$, $i = 1, 2$, is the focus of interest, instead of a complex wave function of a charged particle as in conventional Maxwell theory. In the case of the Sagnac effect, we assign $i = (1)$ to clockwise, and $i = (2)$ to counterclockwise propagating beams, but the clockwise and counterclockwise propagating beams in the interferometric situation we are considering can only be distinguished after parallel transport following a *patch condition*, which is a condition initiated when the Sagnac interferometer platform undergoes an angular rotation. A change in phase then means a change in the orientation in *"internal space"* under the transformation $\psi \rightarrow S\psi$, where S is a unitary matrix with a unitary determinant.

A.
DIRECT PRODUCT OF COUNTERPROPAGATING BEAMS

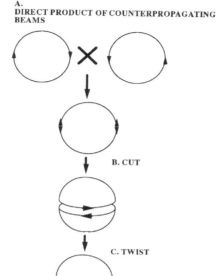

B. CUT

C. TWIST

D. JOIN

Fig. 1.4.3.   Generation of a topological obstruction (source of magnetic flux) in the Sagnac interferometer. Commencing with a direct product of two counterpropagating beams, on S^3, A, the sphere S^3 is cut in B, given a twist in C and joined in D. The twist and join in D corresponds to a "patch" condition, and in the Sagnac interferometer it is caused by angular velocity (+ linear acceleration). Figure 1.4.3(C) corresponds to Figs. 1.4.1(D) and 1.4.2(C); and Fig. 1.4.3(D) to Figs. 1.4.1(E) and 1.4.2(D). (Adapted from Ref. 13.)

With the gauge potential now defined as a matrix, $A_\mu(x)$, with g as a generalized charge, and with parallel transport along separate arms of the interferometer, we have,

beam traveling *clockwise* and platform *rotated* (at x):

$$\psi(x) \rightarrow \psi'(x) = S(x)\psi(x), \qquad (1.4.24a)$$

beam traveling *counterclockwise* and platform *rotated* (at $x + dx^\mu$):

$$\hat{\psi}(x) \rightarrow \hat{\psi}'(x) = S(x + dx^\mu)\hat{\psi}(x), \qquad (1.4.24b)$$

or, more explicitly:

beam traveling *clockwise* and platform *rotated* (at x):

$$\psi_P(x) \to_{\Gamma_{\text{clockwise}}} \psi_Q(x) = \exp[igA_\nu(x + dx^\mu)dx^\nu]$$
$$\times \exp[igA_\mu(x)dx^\mu]\psi_P(x), \qquad (1.4.25a)$$

beam traveling *counterclockwise* and platform *rotated* (at $x + dx^\mu$):

$$\psi_P(x) \to_{\Gamma_{\text{counterclockwise}}} \psi_Q(x) = \exp\left[igA_\mu(x + dx^\nu)dx^\mu\right]$$
$$\times \exp[igA_\nu(x)dx^\nu]\psi_P(x). \qquad (1.4.25b)$$

Taking the difference and using Stokes' theorem gives

$$F_{\mu\nu}(x) = \partial_\nu A_\mu(x) - \partial_\mu A_\nu(x) + ig[A_\mu(x), A_\nu(x)]. \qquad (1.4.26)$$

This is the description we seek. $F_{\mu\nu}(x)$ [Eq. (1.4.26)] is *gauge–covariant* owing to the phase direction in internal symmetry space, which is not the case with $f_{\mu\nu}$ [Eq. (1.4.23)]. The field at H (cf. Fig. 1.1) in the Sagnac interferometer with the platform *under rotation* is then described by

$$F'_{\mu\nu} = S(x)F_{\mu\nu}(x)S'(x), \qquad (1.4.27)$$

where $S(x)$, and its converse, $S'(x)$, constitute a gauge group to be defined.

It should be emphasized that this description applies only to the platform *under rotation*. On cessation of rotation, the Sagnac interferometer platform no longer exhibits a patch condition, whereupon

$$A_\mu \to a_\mu,$$
$$A_\nu \to a_\nu,$$
$$g \to 0, \qquad (1.4.28)$$
$$(\partial_\nu A_\mu(x) - \partial_\mu A_\nu(x) + ig[A_\mu(x), A_\nu(x)]) \to (\partial_\nu a_\mu(x) - \partial_\mu a_\nu(x)),$$
$$F_{\mu\nu} \to f_{\mu\nu},$$

and we recapture conventional Maxwell theory. Upon rerotation of the frame of reference of the platform of the Sagnac interferometer, we again have

$$a_\mu \to A_\mu,$$
$$a_\nu \to A_\nu,$$
$$g > 0, \qquad\qquad (1.4.29)$$
$$(\partial_\nu a_\mu(x) - \partial_\mu a_\nu(x)) \to (\partial_\nu A_\mu(x) - \partial_\mu A_\nu(x) + ig[A_\mu(x), A_\nu(x)]),$$
$$f_{\mu\nu} \to F_{\mu\nu},$$

and we recapture Yang–Mills theory once again. Stated differently, with rotation of the platform, the gauge symmetry is SU(2)/$Z_2 \cong$ SO(3), and on stabilization of the platform the gauge symmetry is U(1).[o]

2.   The Lorentz Group and the Lorenz Gauge Condition

The Lorentz group expresses a space–time symmetry in which free space exhibits the same physical properties in all inertial frames. Accelerated frames are deviations from inertial frames and the acceleration can be gravitational or kinematic in origin. Gyroscopes measure deviation of frames from an inertial frame. Therefore, rotation of the Sagnac interferometer platform causes a deviation from the Lorentz group description of space–time.

If W is a vector defined over the vector space of real numbers and in the Minkowskian metric, then a *Lorentz norm*[p] is defined as

$$\| W \| = W \cdot W = W^i W^j \eta_{ij} = (W^0)^2 + (W^1)^2 + (W^2)^2 + (W^3)^2. \quad (2.1)$$

If V is a vector similarly defined, then[26]

$$W \cdot V = \frac{1}{2}\{\| W + V \| - \| W \| - \| V \|\} \qquad (2.2)$$

is the *inner product* defined in terms of the Lorentz norm. Therefore, a *Lorentz transformation* on W in vector space can be defined as preserving the Lorentz norm, the inner product, the spatial orientation and the time orientation of W.

---

[o]Gauge transformations of U(1) are obtained by means of multiplication by complex scalar fields of unit modulus — or by fields of elements of the Lie group U(1) (cf. Ref. 26, p. 342).
[p]H.A. Lorentz (1853–1928) of Leiden.

If S and T are points in the Minkowski metric and ST is the position vector of T relative to S, then

$$\chi(S, T) = \|ST\| \qquad (2.3)$$

is the *squared interval* for any pair of points S and T in Minkowski space. A transformation of Minkowski space which preserves the squared interval is a *Poincaré transformation*.

The Lorenz[q] gauge condition is

$$\nabla^a a_a = 0, \qquad (2.4)$$

which guarantees invariance of the Maxwell equations under transformations. The Lorenz gauge is conformally invariant. However, the Maxwell equations with the Lorenz gauge *are not conformally invariant owing to a conformal invariance mismatch* (cf. Ref. 26, p. 373). Yet this is a minor point. The major one is that the Lorenz gauge and the Lorentz group are not necessary choices *for all platform conditions*. According to the present analysis, they are also *incorrect choices* when the Sagnac interferometer platform is *in rotation*.

3. The Phase Factor Concept

In conventional electromagnetism, on the one hand, $f_{\mu\nu}$ *underdescribes* the system because phase is undetermined; on the other hand, $a_\mu(x)$ *overdescribes* the system because different values of $a_\mu(x)$ correspond to the same physical condition.[54] But the Dirac phase factor defined as

$$\Phi(C) = \exp\left[ie \oint_C a_\mu(x)dx^\mu\right], \qquad (3.1)$$

where e is electric charge and a_μ is the electromagnetic potential, *completely* describes the system in the case of the Sagnac interferometer fields when the interferometer platform is *at rest*.

For the case of the Sagnac interferometer in rotation and multiply connected in a generated A_μ matrix field of specific symmetry, Φ may

---

[q]L. Lorenz (1829–1891) of Copenhagen.

be said to be in an excited state, Φ^*. Accordingly, in Yang–Mills theory,

$$\Phi^*(C) = P \exp\left[ig \oint_C A_\mu(x)dx^\mu\right], \qquad (3.2)$$

where $\Phi^*(C)$ specifies parallel phase transport over any loop C in rotation, g is generalized charge, P specifies path dependence of the integral and A_μ is a matrix variable. As $F_{\mu\nu}$ specifies only infinitesimal loops (i.e. local conditions), $\Phi^*(C)$ is the global version of $F_{\mu\nu}$ and differs from $F_{\mu\nu}$ *if the region is multiply connected.* The phase factor, $\Phi^*(C)$, describes the Sagnac interferometer fields when the interferometer platform is *rotated.*

One may then ask: if the surface of the Sagnac interferometer, Σ, is continuously deformed in space, do the paths Γ_1 and Γ_2 deform to a point in a gauge group? Our answer is: *yes,* if the platform of the interferometer is *not* rotated, but *no,* if the platform of the interferometer *is* rotated. The counterpropagation of the two beams around an obstruction permits deformation to a point only in the nonrotated condition. When rotated, a "patch" condition (cf. Figs. 1.4.1–1.4.3) exists in the multiply connected topology. The measured phase difference of the rotated Sagnac interferometer is then

$$\oint \left(A_\mu^{\text{clockwise}} - A_\mu^{\text{anticlockwise}}\right)dx^\mu = \oint (\Delta\phi)\,dx^\mu = 4\pi g_m, \qquad (3.3)$$

where g_m is "magnetic charge." The counterpropagating beams on a rotating platform are an instance of a patching condition which precludes shrinkage to a single point. Therefore Eq. (3.3) describes the dynamic explaining the Sagnac effect. On the other hand, the *unrotated* Sagnac interferometer has no measured phase difference and for that stabilized state

$$\oint \left(a_\mu^{\text{clockwise}} - a_\mu^{\text{anticlockwise}}\right)dx^\mu = 0. \qquad (3.4)$$

Whereas in conventional electromagnetics a_μ and $f_{\mu\nu}$ are labeled only at *points* in space–time, in gauge theory $\Phi(C)$ and $\Phi^*(C)$ are labeled by C, a *closed curve* in space–time. $\Phi^*(C)$ is a functional of the function

$$A = \{A^\mu(s); s = 0 \to 2\pi\}, \tag{3.5}$$

and the relation to the field tensor, F_μ, is

$$F_\mu(A[s]) = \Phi_A^{*-1}(s, 0) F_{\mu\nu}(A(s)) \Phi_A^*(s, 0) \frac{dA^\nu(s)}{ds}. \tag{3.6}$$

3.1. *SU(2) group algebra*

The explanation we have sought for the Sagnac effect pivots on an understanding of the SU(2) continuous group. The following is an account of how that group relates to the U(1) group of rotations.

A general rotation about some arbitrary axis is designated $R(\alpha, \beta, \gamma)$ in group O(3) for the Euler angles α, β and γ. The SU(2) group has the following group elements which refer to a complex two-dimensional vector (u, v):

$$\begin{array}{c} u' \\ v' \end{array} = \begin{bmatrix} a & b \\ c & d \end{bmatrix} \begin{array}{c} u \\ v \end{array}, \tag{3.1.1}$$

where a, b, c and d are complex numbers. With the additional requirement that the determinant be ± 1, making the group unitary, and $ad - bc = +1$, making the group special (S) (so special unitary is a subclass of the unitary group), the transformation rules simplify to the matrix

$$\begin{bmatrix} a & b \\ -b^* & a^* \end{bmatrix}, \tag{3.1.2}$$

which is the defining matrix for the SU(2) group of continuous transformations. In terms of the $R(\alpha, \beta, \gamma)$ rotation, and as an example, we could choose $a = \exp[\frac{-i\alpha}{2}]$ and $b = 0$, which gives a rotation $R(\alpha, 0, 0)$ about the z axis, and $a = \cos[\frac{\beta}{2}]$ and $b = \sin[\frac{\beta}{2}]$, which gives the rotation $R(0, \beta, 0)$ about the y axis. Then rotations $R(\alpha, \beta, \gamma)$ would then be associated with the SU(2) matrix:

$$
\begin{bmatrix}
\cos\left[\frac{\beta}{2}\right]\exp\left[\frac{i(\alpha+\gamma)}{2}\right] & \sin\left[\frac{\beta}{2}\right]\exp\left[\frac{-i(\alpha-\gamma)}{2}\right] \\
-\sin\left[\frac{\beta}{2}\right]\exp\left[\frac{i(\alpha-\gamma)}{2}\right] & \cos\left[\frac{\beta}{2}\right]\exp\left[\frac{-i(\alpha+\gamma)}{2}\right]
\end{bmatrix}.
\tag{3.1.3}
$$

These SU(2) transformations define the relations between the Euler angles of group O(3) with the parameters of SU(2). Each of two fields, $(u, v) \rightarrow (u', v')$, can be associated with the rotation matrix in O(3) for $R(\alpha, \beta, \gamma)$:

$$
\begin{array}{ccc}
\cos(\alpha)\cos(\beta)\cos(\gamma) - \sin(\alpha)\sin(\beta) & \sin(\alpha)\cos(\beta)\cos(\gamma) + \cos(\alpha)\sin(\gamma) & -\sin(\beta)\cos(\gamma) \\
-\cos(\alpha)\cos(\beta)\sin(\gamma) - \sin(\alpha)\cos(\gamma) & -\sin(\alpha)\cos(\beta)\sin(\gamma) + \cos(\alpha)\cos(\gamma) & \sin(\beta)\sin(\gamma) \\
\cos(\alpha)\sin(\beta) & \sin(\alpha)\sin(\beta) & \cos(\beta)
\end{array}
$$

$$\tag{3.1.4}$$

Then matrix (3.1.2) defining SU(2) algebra can be related to O(3) with algebra defined by matrix (3.1.4). This relationship is called a *homomorphism*.

The SU(2) group is a Lie algebra such that for the angular momentum generators, J_i, the commutation relations are $[J_i, J_j] = i\varepsilon_{ijk}J_k$, $i, j, k = 1, 2, 3$, where ε_{ijk} are "structure constants." The four dimensions *can be the three Euclidean spatial dimensions and time (or Minkowski space–time), but need not be. Here, they are related to a complex space–time with a holomorphic metric* (see below).

As an example of an SU(2) transformation, we show the following. An isotropic parameter, w, can be defined:

$$
w = \frac{x - iy}{z},
\tag{3.1.5}
$$

where x, y, z are the spatial coordinates. If w is written as the quotient of μ_1 and μ_2, or the homogeneous coordinates of the bilinear transformation, then, corresponding to (3.3) and using matrix (3.1.3) we have the following example of a vector undergoing an SU(2) transformation:

$$
|\mu_1'\mu_2'\rangle =
\begin{bmatrix}
\cos\left[\frac{\beta}{2}\right]\exp\left[\frac{i(\alpha+\gamma)}{2}\right] & \sin\left[\frac{\beta}{2}\right]\exp\left[\frac{-i(\alpha-\gamma)}{2}\right] \\
-\sin\left[\frac{\beta}{2}\right]\exp\left[\frac{i(\alpha-\gamma)}{2}\right] & \cos\left[\frac{\beta}{2}\right]\exp\left[\frac{-i(\alpha+\gamma)}{2}\right]
\end{bmatrix}
|\mu_1\mu_2\rangle.
$$

$$\tag{3.1.6}$$

The *sourceless* condition for such transformations is that the action

$$L = \oiiint f^i_{\mu\nu} f^i_{\nu\mu} dS = \text{an absolute minimum,} \qquad (3.1.7)$$

where the integral is taken over a closed four-dimensional surface. The solutions to this equation are gauge fields (nonintegrable phase factors) implying conformal invariance.

Let us define further restrictions or boundary conditions. If we define the coordinates:

$$x = \frac{1}{2}(u^2 - v^2),$$

$$y = \frac{1}{2i}(u^2 + v^2), \qquad (3.1.8)$$

$$z = uv,$$

then

$$x^2 + y^2 + z^2 \text{ is invariant.} \qquad (3.1.9)$$

Suppose that we let $\alpha = \gamma = 0$ and choose $\beta = 0$ or $\beta = 2\pi$. For $\beta = 0$ the SU(2) matrix is

$$\begin{bmatrix} 1 & 0 \\ 0 & 1 \end{bmatrix}. \qquad (3.1.10)$$

However, for $\beta = 2\pi$ the SU(2) matrix is

$$\begin{bmatrix} -1 & 0 \\ 0 & -1 \end{bmatrix}. \qquad (3.1.11)$$

Therefore, for zero rotation in three-dimensional space, there correspond *two distinct* SU(2) elements depending on the value of β. The two-to-one relationship of a U(1) group to an SU(2) form is an indication of compactification of degrees of freedom.

Previously,[2,3] the conventional U(1) symmetry Maxwell theory was placed in a Yang–Mills context and generalized to an SU(2) symmetry form. Table 3.1.1 compares the two formulations.

Table 3.1.1.

U(1)	SU(2) Symmetry
$\rho_e = J_0$	$\rho_e = J_0 - iq(A \cdot E - E \cdot A) = J_0 + qJ_z$
$\rho_m = 0$	$\rho_m = -iq(A \cdot B - B \cdot A) = -iqJ_y$
$g_e = J$	$g_e = iq[A_0, E] - iq(A \times B - B \times A) + J$
	$\quad = iq[A_0, E] - iqJ_x + J$
$g_m = 0$	$g_m = iq[A_0, B] - iq(A \times E - E \times A) = iq[A_0, B] - iqJ_z$
$\sigma = J/E$	$\sigma = \dfrac{\{iq[A_0, E] - iq(A \times B - B \times A) + J\}}{E}$
	$\quad = \dfrac{\{iq[A_0, E] - iqJ_x + J\}}{E}$
$s = 0$	$s = \dfrac{\{iq[A_0, B] - iq(A \times E - E \times A)\}}{H}$
	$\quad = \dfrac{\{iq[A_0, B] - iqJ_z\}}{H}$

where the Noether currents are

$$J_x = (A \times B - B \times A),$$
$$J_y = (A \cdot B - B \cdot A),$$
$$J_z = (A \times E - E \times A),$$
$$iJ_z = (A \cdot E - E \cdot A).$$

(3.1.12)

Table 3.1.2 compares the Maxwell equations formulated for U(1) group symmetries, i.e. the conventional form, and the Maxwell equations formulated for SU(2) group symmetries.

Table 3.1.2.   Maxwell Equations

	U(1)	SU(2)
Gauss's law	$\nabla \cdot E = J_0$	$\nabla \cdot E = J_0 - iq(A \cdot E - E \cdot A)$
Ampère's law	$\dfrac{\partial E}{\partial t} - \nabla \times B + J = 0$	$\dfrac{\partial E}{\partial t} - \nabla \times B + J + iq[A_0, E]$
		$\quad -iq(A \times B - B \times A) = 0$
Coulomb's law	$\nabla \cdot B = 0$	$\nabla \cdot B + iq(A \cdot B - B \cdot A) = 0$
Faraday's law	$\nabla \times E + \dfrac{\partial B}{\partial t} = 0$	$\nabla \times E + \dfrac{\partial B}{\partial t} + iq[A_0, B]$
		$\quad +iq(A \times E - E \times A) = 0$

3.2. *A short primer on topological concepts*

The present essay attempts to explain a physical effect, the Sagnac effect, using topological concepts *as necessary descriptive forms*, besides using well-known mathematical analysis techniques. Such topological concepts are not commonly in the tool kit of most physicists and engineers. It is a fact of human experience that without the correct mathematical "filter," a physical problem cannot be correctly defined. A major reason for the use of topology is that it is a study of continuity and continuous deformation, but, above all, it provides *justification for the existence or nonexistence of a qualitative, rather than a quantitative, object*. The conventional mathematical tool kit is generally one for quantitative prediction and *such methods already assume the existence of the qualitative objects they address*. The present approach proposes that the correct mathematical filter or descriptive tool is a topological filter or a topological tool kit (see the following chapter). That is, the physical problem offered by the Sagnac effect cannot be defined without topological concepts. Therefore, the following section is a very short introduction to definitions of some useful topological concepts.

Topological invariants
In a topological space, X, the number of connected parts and the number of holes are topological invariants if the number does not change under homomorphism. Topological invariants uniquely specify equivalence classes.

Homomorphism
If $\phi_x : X \rightarrow X'$ is continuous and if ϕ_x^{-1} exists and is also continuous, so that X and X' have the same number of components and holes, then ϕ_x is a homomorphism and a *coordinate mapping* and X is a *coordinate chart*.

Manifold or differentiable manifold
A manifold of real dimension is a topological space which is locally homomorphic to an open set of the real numbers. A differential manifold is a manifold with the property that if X and Y are coordinate charts which have a nonempty intersection, then the mapping $\phi_X \circ \phi_Y$

is a differential mapping. In other words, a differentiable manifold is the primitive topological space for the study of differentiability.

Homotopy classes

If Γ defines closed curves on a gauge group G, if Γ cannot be shrunk to a point and if members of a class of Γ cannot be continuously deformed into each other, then that class is the homotopy class of Γ. Stated differently, two coterminous paths in parameter space are *homotopic* if they can be continuously deformed into each other. Thus, whereas homomorphisms define equivalence classes of topological *spaces*, homotopy defines equivalence classes of continuous *maps*.

Connectedness

If

(1) the zeroth homotopy set of X, or $\Pi_0(X)$, is the set of path-connected components of X,
(2) $\Pi_1(X)$ is the set of loops which cannot be continuously deformed into each other, and
(3) $\Pi_1(X) = 0$,
  then the space X is *simply connected*.

If $\Pi_1(X) \neq 0$, then the space X is said to be *multiply connected*.

If $\Pi_n(X) = \Pi_{n-1}(\Omega X), n \geq 1$, and for $n \geq 1$, $\Pi_n(X)$ is a *group*, but $\Pi_0(X)$ is not a group.

If X is a Lie group, then $\Pi_0(X)$ inherits a group structure from X, because it can be identified with the quotient group of X by its identity-connected component. A *connection* can always be defined independently of the choice of metric on the space.

Topological obstruction

If X and X' are two equivalence classes, and X cannot be deformed continuously into X', then there exists a topological obstruction preventing a mapping, which is a topological invariant of X.

Fiber bundles

A fiber bundle is a twisted product of two spaces, X and Y, where Y is acted on by a group G, and the twist in the product has been effected by the group action. X is the base space and Y is the *fiber* on which the group structure acts. A rotating Sagnac interferometer is thus a nontrivial fiber bundle over X with *total space* or *topological space E* such that $\Pi : E \to X$. In that interferometric situation, X, the base space is identified with the Sagnac interferometer *with platform at rest* and E is identified with the Sagnac interfometer *with platform in motion*. The clockwise and counterclockwise propagating beams play the role of two disjoint arcs on X. With the Sagnac interferometer at rest, coordinates in A are equivalent to coordinates in B. However, for the Sagnac interferometer platform in motion, changing coordinates in A to coordinates in B (or vice versa) requires a transfer function or group element G. For the platform in motion, but not at rest, the fiber Y is acted on by the group element G. The fiber Y in the Sagnac interferometer is the local propagating beam segment for both clockwise and counterclockwise beams and common to the Sagnac interferometer at rest and in motion.

Holomorphic

A differential function defined on an open set of complex numbers is said to be holomorphic if it satisfies the Cauchy–Rieman equations. A holomorphic transformation is a complex analytic transformation for which the real and imaginary parts are Taylor-expandable. It is also conformal and orientation-preserving.

Conformal

A conformal structure on a manifold is the prescription of a null cone defined by a quadratic function in the tangent space at each point of the manifold. A conformal mapping is an angle-preserving mapping.

Conformal structure

The concept of *conformal structure* is related to that of *conformal rescaling*. In the case of conventional electromagnetism, significance is only given to an equivalence class of fields which can be obtained from a given metric, g_{ab}, by a conformal rescaling:

$$g_{ab} \mapsto \hat{g}_{ab} = \Omega^2 g_{ab}, \qquad (3.2.1)$$

where Ω is any scalar field. No transformation of *points* is involved, *but information is lost when a field undergoes conformal rescaling in conventional electromagnetism.* However, if a spinor description is used, we have

$$\varepsilon_{AB} \mapsto \hat{\varepsilon}_{AB} = \Omega \varepsilon_{AB}. \tag{3.2.2}$$

Therefore, whereas the argument of the inner product between two spin vectors is conformally invariant, *the modulus of the inner product is altered under conformal rescaling.* Maxwell's equations of conventional electromagnetism are conformally invariant, and so is the Lorenz gauge. However, Maxwell's equations with the Lorenz gauge are not conformally invariant (cf. Ref. 26, p. 373). The Sagnac interferometer exhibits conformal invariance under rotation.

Conformal invariance

A system of fields and field equations is conformally invariant if it is possible to attach conformal weights to all field quantities in such a way that the field equations remain true after conformal rescaling. The Euler–Lagrange equations

$$D^*F = 0 \tag{3.2.3}$$

(where D is the covariant exterior derivative), the self-dual and the anti-self-dual equations

$$F = \mp^* F \tag{3.2.4}$$

for the two-form F, and the action

$$S = \frac{-1}{16} \int d^4 x f_{\mu\nu}(x) f^{\mu\nu}(x) \tag{3.2.5}$$

are all conformally invariant.

Holonomy

The holonomy of a given closed curve is the parallel transport considered as an element of the structure group under an embedding.

Cohomology

Cohomology groups are dual to homology groups. A homology group can be formed from a class C of p chains, which is a formal finite linear combination. If C_p is the set of all $C^\infty p$ chains, then ω is a map from C_p to R

$$\omega : C_p \to R, \qquad (3.2.6)$$

or to an element of the dual of C_p. Therefore, if $H_p(M; Z)$ is a homology group, $H^p(M; R)$ is the corresponding cohomology group, where the de Rham cohomology group[9] is

$$H^p = \{\text{closed } p \text{ form}\}/\{\text{exact } p \text{ form}\}.$$

In the case of Stokes' theorem,

$$\int_M D\omega = \int_{\partial M} \omega, \qquad (3.2.7)$$

where ∂ is the boundary operator and D is the exterior derivative. If

$$\langle \omega, C \rangle = \int_C \omega, C \in C_p, \qquad (3.2.8)$$

then Stokes' theorem is

$$\langle D\omega, C \rangle = \langle \omega, \partial C \rangle . \qquad (3.2.9)$$

de Rham's theorem is (Ref. 45, 1990, p. 161)

$$H^q(M, C) \cong H^q_{dr}(M), \qquad (3.2.10)$$

or, in words, the complex cohomology group, $H^q(M, C)$, is equivalent to the qth de Rham cohomology group of the manifold M, $H^q_{dr}(M)$.

Differential forms

In Euclidean space R^n the coordinates $x^1, \ldots, x^n$ are O *forms* and differentiation is the gradient. In R^n the differentiation on covariant vector fields, e.g. $A_\mu dx^\mu$ on R^n, is the curl and those fields are *one-forms*. So the differential symbol can represent a linear map from O forms to one-forms and correspond to line integrals.

In R^3 the area $A_{ij}dx^i dx^j$ for a two-dimensional surface is a *two-form* and differentiation then is an exterior differentiation or divergence.

For every form ξ in a Riemannian manifold, there is a dual form $^*\xi$:

$$^{*\cdot}\xi^P \rightarrow \xi^{n-P}, \text{ e.g.,} \tag{3.2.11}$$

$$^{*\cdot} f_{\mu\nu} \rightarrow -\frac{1}{2}\varepsilon_{\mu\nu\rho\sigma} f^{\rho\sigma}, \tag{3.2.12}$$

where f is the electromagnetic tensor. If $d\xi = 0$, ξ is said to be *closed*. If $\xi = df$, where f is any function, ξ is said to be *exact*.

Self-duality and anti-self-duality

If we commence with the spinor form of the Yang–Mills field tensor, we have[r]

$$F_{ab} = \varphi_{AB}\varepsilon_{A'B'} + \varepsilon_{AB}\bar{\varphi}_{A'B'}. \tag{3.2.13}$$

The *dual* of F_{ab}, $^*F_{ab}$, is

$$^*F_{ab} = -i\varphi_{AB}\varepsilon_{A'B'} + i\varepsilon_{AB}\psi_{A'B'}, \tag{3.2.14}$$

where[s]

$$\varphi_{AB} = \varphi_{(AB)} = \frac{1}{2}F_{ABC'}^{C'}, \tag{3.2.15}$$

$$\psi_{A'B'} = \psi_{(A'B')} = \frac{1}{2}F_{C}^{CA'B'}. \tag{3.2.16}$$

---

[r]In this equation: (1) a prime on a label indicates complex conjugation; (2) the ε spinor is used, which is *antisymmetrical*: $\varepsilon_{AB} = -\varepsilon_{BA}$ or $\varepsilon_A^B = -\varepsilon_A^B$ and $x_B = x^A\varepsilon_{AB}$; $x^A = \varepsilon^{AB}x_B$; and (3) a bar on a spinor indicates complex conjugation; (4) φ_{AB} is a *symmetric* spinor; (5) the indices AB are abstract indices, without any reference to any basis or coordinate system; and (6) the clumped pair of indices are defined as $a = AA', b = BB', c = CC', \ldots, z = ZZ'^{26}$.

[s]Round brackets indicate symmetrization and square brackets indicate antisymmetrization, e.g.

$$U_{A(BC)D} = \frac{1}{2!}(U_{ABCD} + U_{ACBD}),$$

$$U_{A(BCD)E} = \frac{1}{3!}(U_{ABCDE} + U_{ACBDE} + U_{ACDBE} + U_{ABDCE} + U_{ADBCE} + U_{ADCB}),$$

$$U_{A[BC]D} = \frac{1}{2!}(U_{ABCD} - U_{ACBD}),$$

$$U_{A[BCD]E} = \frac{1}{3!}(U_{ABCDE} - U_{ACBDE} + U_{ACDBE} - U_{ABDCE} + U_{ADBCE} - U_{ADCB}).$$

If

$$^*F_{ab} = iF_{ab}, \text{ i.e. } \varphi_{AB} = 0,^{\text{t}} \tag{3.2.17}$$

the bivector field F_{ab} is *self-dual*. If

$$^*F_{ab} = -iF_{ab}, \text{ i.e. } \psi_{A'B'} = 0, \tag{3.2.18}$$

the bivector field F_{ab} is *anti-self-dual*.

In other words, every complex bivector F_{ab} is a sum of self-dual, $^+F_{ab}$, and anti-self-dual, $^-F_{ab}$, components:

$$F_{ab} = {}^+F_{ab} + {}^-F_{ab}, \tag{3.2.19}$$

where

$$^+F_{ab} = \frac{1}{2}(F_{ab} - i^*F_{ab}) = \varepsilon_{AB}\psi_{A'B'}, \tag{3.2.20}$$

$$^-F_{ab} = \frac{1}{2}(F_{ab} + i^*F_{ab}) = \varphi_{AB}\varepsilon_{A'B'}. \tag{3.2.21}$$

In the case of Yang–Mills fields we have[u]

$$^+F_{ab\Theta}^{\Psi} = \varepsilon_{AB}\chi_{A'B'\Theta}^{\Psi}, \tag{3.2.22}$$

$$^-F_{ab\Theta}^{\Psi} = \varphi_{AB\Theta}^{\Psi}\varepsilon_{A'B'}. \tag{3.2.23}$$

Affine transformation

If A is a connection form, and if the local coordinates are changed,

$$(x, g) \to (x', g'); \quad g' = hg, \tag{3.2.24}$$

then

$$A' = hdh^{-1} + hAh^{-1} \tag{3.2.25}$$

is an affine transformation. It is affine because it both translates (by the amount hdh^{-1}) and rotates A (according to hAh^{-1}). Owing to

---

[t]The clumped pairs of spinor indices are defined as
$a = AA', b = BB', c = CC', \ldots, z = ZZ'$. Therefore
$$\psi_{BB'}^{AA'\,Q} = \psi_{BB'}^{a\,\ Q} = \psi_b^{a\,Q} = \psi_b^{A'A\,Q} = \psi_{BB'}^{a\,\ Q},$$
$$\psi_{AA'}^{a\,\ \ Q} = \psi_a^{AA'\,Q} = \psi_a^{a\,Q} = \psi_{AA'}^{AA'\,Q},$$
$$\psi_{BB'}^{AA'\,B} = \psi_b^{aB} = \psi_b^{AA'\,B} = -\psi_b^{aB} = -\psi_{BB'}^{aB}$$
(Ref. 26, p. 116).
[u]Capital Greek labels are used to indicate elements of vector spaces. That is, they are bundle indices. This relabeling indicates that a component λ^Φ *locally* assigned to a Yang–Mills field has a basis which corresponds to that of the Yang–Mills field in the *global* manifold. The set of fields α_Ψ^Ψ is a gauge for τ^Ψ, the tensor describing a ring of scalar fields. (Ref. 26, p. 345.)

the presence of the translational quantity, **A** cannot be represented in tensorial form (Ref. 24, p. 178).

Action invariance and gauge transformations

Because of the assumption of local gauge invariance, the action is invariant under local gauge transformations. The Euler–Lagrange equations and the action are conformally invariant. The dual operator ∗ is also conformally invariant.

4.   Minkowski Space–Time Versus Cartan–Weyl Form

Minkowski space, or vector space, is the space–time of *special relativity*. In the curved space–time of *general relativity*, Minkowski vector spaces occur as the *tangent spaces* of space–time events. Minkowski space is a four-dimensional vector space over the field of real numbers possessing (a) an orientation, (2) a bilinear inner product of signature (+ − − −) and (3) a time orientation. The signature refers to a tetrad — or four linearly independent vectors — t, x, y, z, such that

$$t \cdot t = 1,$$
$$x \cdot x = y \cdot y = z \cdot z = -1, \tag{4.1}$$
$$t \cdot x = t \cdot y = t \cdot z = x \cdot y = x \cdot z = y \cdot z = 0.$$

If

$$t = g_0, x = g_1, y = g_2, z = g_3, \tag{4.2}$$

then

$$g_i \cdot g_j = \eta_{ij}, \tag{4.3}$$

where

$$(\eta_{ij}) = (\eta^{ij}) = \begin{pmatrix} 1 & 0 & 0 & 0 \\ 0 & -1 & 0 & 0 \\ 0 & 0 & -1 & 0 \\ 0 & 0 & 0 & -1 \end{pmatrix}. \tag{4.4}$$

The Minkowski space–time or group refers to a well-defined metric and is the underlying algebraic logic for the special theory of relativity. However, as is well known, the special theory of relativity is only valid locally in a frame in free fall, which disconnects it from any gravitational effects (Ref. 15, p. 46) — hence the reason for the appellative "special."

Make a comparison now with the Weyl group. The Weyl group has a well-defined conformal structure but no preferred metric. The Riemann tensor of general relativity has two components: the Ricci tensor and the Weyl tensor.[47] The Ricci tensor reflects the distribution of matter fields and the Weyl tensor describes space curvature which is not *locally* determined by the matter density. In terms of the *Weyl conformal tensor*, C^d_{abc},[47] the *Weyl tensor in spinor form*, Ψ_{ABCD}, is[v]

$$C_{AA'BB'CC'DD'} = \Psi_{ABCD}\varepsilon_{A'B'}\varepsilon_{C'D'} + \varepsilon_{AB}\varepsilon_{CD}\bar{\Psi}_{A'B'C'D'}. \tag{4.5}$$

where $\bar{\Psi}_{A'B'C'D'}$ is the complex conjugate of the four-valent spinor Ψ_{ABCD}.

In a space–time with Lorentzian signature, the self-dual and the anti-self-dual parts of the Weyl tensor (namely C^+_{abcd} and C^-_{abcd}), or of the Weyl spinor ($\bar{\Psi}_{A'B'C'D'}$ and Ψ_{ABCD}), are complex conjugates of each other.[w] Therefore, an anti-self-dual space–time (one with $C^+_{abcd} = 0$) is necessarily conformally flat ($C_{abcd} = 0$). However, this restriction does not apply to positive-definite four-spaces, or to complex space–times (Ref. 45, p. 293). By a "complex space–time" is meant a four-dimensional complex manifold M, equipped with a *holomorphic* metric g_{ab}. In other words, with respect to a holomorphic coordinate basis $x^a = (x^0, x^1, x^2, x^3)$ the metric is a 4×4 matrix of holomorphic functions of x^a, and its determinant is nowhere vanishing. The Ricci tensor (which before was real) becomes complex-valued and the self-dual and

---

[v]As before, in the following equation: (1) a prime on a label indicates complex conjugation, so that spinors come in pairs; (2) the ε spinor is used, which is *antisymmetrical* — $\varepsilon_{AB} = -\varepsilon_{BA}$ or $\varepsilon^B_A = -\varepsilon^B_A$ and $x_B = x^A \varepsilon_{AB}$; $x^A = \varepsilon^{AB} x_B$, i.e. the ε spinors are used to raise and lower indices on other spinors; (3) a bar on a spinor indicates complex conjugation; and (4) the indices AB are abstract indices, without any reference to any basis or coordinate system.[26]

[w]As before, the clumped pairs of spinor indices are defined as

$a = AA'; b = BB', c = CC', \dots, z = ZZ'$. Therefore

$$\psi^{AA'Q}_{BB'} = \psi^{a\,Q}_{BB'} = \psi^{a\,Q}_b = \psi^{A'AQ}_b = \psi^{a\,Q}_{BB'},$$

$$\psi^{a\,Q}_{AA'} = \psi^{AA'Q}_a = \psi^{a\,Q}_a = \psi^{AA'Q}_{AA'},$$

$$\psi^{AA'B}_{BB'} = \psi^{aB}_b = \psi^{AA'B}_b = -\psi^{aB}_b = -\psi^{aB}_{BB'}$$

(Ref. 26, p. 116).

anti-self-dual parts of the Weyl tensor (which before were complex-valued but conjugate to each other) become *independent* holomorphic tensors. The vanishing of the Weyl tensor implies that the space–time is locally Minkowskian.

The origins of the Weyl group extend back to the discovery of spinors by Cartan in 1913.[6,7] Although the development of spinors has proceeded formally,[5,49,50] the original Cartan approach provides geometrical definitions of the mathematical entities.

Whatever the global conditions of this complex space–time, the following Riemann manifolds are possible[45]: (1) M is compact, without boundary; (2) (M, g_{ab}) is asymptotically locally Euclidean (ALE); and (3) (M, g_{ab}) is asymptotically locally flat (ALF). In empty space the Ricci tensor vanishes, and the Bianchi identity equation is

$$\nabla^{AA'}\Psi_{ABCD} = 0, \tag{4.6}$$

which is of the same form as the spinor description of the Maxwell equations:

$$\nabla^{AA'}\varphi_{AB} = 0. \tag{4.7}$$

This symmetric spinor object defines the Maxwell field tensor,[x]

$$f_{AA'BB'} = \varphi_{AB}\varepsilon_{A'B'} + \varepsilon_{AB}\bar{\varphi}_{A'B'}, \tag{4.8}$$

which, according to our account above, describes the Sagnac interferometer *when the platform is at rest.*

The electromagnetic potential, *a*, can be defined in terms of the covariant derivative as

$$\nabla_a \equiv \partial_a - iea_a, \tag{4.9}$$

where ∇_a is a covariant derivative, ∂_a is the flat space covariant derivative, and e is the charge of the field; or as

$$a_a \equiv \frac{i}{\varepsilon\alpha}\nabla_\alpha\alpha, \tag{4.10}$$

---

[x] φ_{AB} is a *symmetric* spinor.

where α is a gauge, and the charge $e = n\varepsilon$; or as[y]

$$\varphi_{AB} = \nabla_{A'(A}a_{B)}^{A'} = \frac{1}{2}\left(\nabla_{A'A}a_B^{A'} + \nabla_{A'B}a_A^{A'}\right). \qquad (4.11)$$

Using

$$f_{AB} \equiv \frac{i}{\varepsilon\alpha}\Delta_{AB}\alpha, \qquad (4.12)$$

where the commutator is defined as

$$\Delta_{AB} = \nabla_A\nabla_B - \nabla_B\nabla_A = 2\nabla_{[A}\nabla_{B]}, \qquad (4.13)$$

we can then obtain

$$f_{AB} = \nabla_A a_B - \nabla_B a_A, \qquad (4.14)$$

which, in terms of the flat space covariant derivative, recaptures Eq. (1.4.23)

$$f_{AB} = \partial_B a_A(x) - \partial_A a_B(x), \qquad (1.4.23\ \&\ 4.15)$$

again describing the Sagnac interferometer with *platform at rest*.

In the case of the Sagnac interferometer with *platform in rotation*, and using again Yang–Mills fields, the matrix potential, A, is defined as a matrix of covectors,[z]

$$A_{a\Psi}^{\Theta} = i\alpha_{\Psi}^{\Theta}\nabla_a\alpha_{\Psi}^{\Psi}, \qquad (4.16)$$

---

[y]As before, round brackets indicate symmetrization and square brackets indicate antisymmetrization, e.g.

$$U_{A(BC)D} = \frac{1}{2!}(U_{ABCD} + U_{ACBD});$$

$$U_{A(BCD)E} = \frac{1}{3!}(U_{ABCDE} + U_{ACBDE} + U_{ACDBE} + U_{ABDCE} + U_{ADBCE} + U_{ADCB});$$

$$U_{A[BC]D} = \frac{1}{2!}(U_{ABCD} - U_{ACBD});$$

$$U_{A[BCD]E} = \frac{1}{3!}(U_{ABCDE} - U_{ACBDE} + U_{ACDBE} - U_{ABDCE} + U_{ADBCE} - U_{ADCB}).$$

[z]Capital Greek labels are used to indicate elements of vector spaces. That is, they are bundle indices. This relabeling indicates that a component λ^{Φ} *locally* assigned to a Yang–Mills field has a basis which corresponds to that of the Yang–Mills field in the *global* manifold. The set of fields α_{Ψ}^{Ψ} is a gauge for τ^{Ψ}, the tensor describing a ring of scalar fields. (Ref. 26, p. 345.)

and the Yang–Mills field, with obstruction, is defined in spinor form[aa] as (Ref. 27, p. 34)

$$F_{ab\Theta}^{\Psi} = \varphi_{AB\Theta}^{\Psi}\varepsilon_{A'B'} + \varepsilon_{AB}\chi_{A'B'\Theta}^{\Psi}, \qquad (4.17)$$

where

$$\varphi_{AB\Theta}^{\Psi} = \varphi_{(AB)\Theta}^{\Psi} = \frac{1}{2}F_{ABC'\Theta}^{C'}{}^{\Psi} \qquad (4.18)$$

is the Yang–Mills potential and

$$\chi_{A'B'\Theta}^{\Psi} = \chi_{(A'B')\Theta}^{\Psi} = \frac{1}{2}F_{CA'B'\Theta}^{C\Psi}; \qquad (4.19)$$

and in vector potential form as (Ref. 26, p. 349)

$$\frac{1}{2}F_{ab\Theta}^{\Psi} = \nabla_{[a}A_{b]\Theta}^{\Psi} - iA_{\Lambda[a}^{\Psi}A_{b]\Theta}^{\Lambda} = \frac{1}{2!}(\nabla_a A_{b\Theta}^{\Psi} - \nabla_b A_{a\Theta}^{\Psi} - i[A_{a\Lambda}^{\Psi}, A_{b\Theta}^{\Lambda}]),$$
$$(4.20)$$

which, in terms of the flat space covariant derivative and, but for a generalized charge $-g$, recaptures Eq. (1.4.26):

$$F_{\mu\nu}(x) = \partial_\nu A_\mu(x) - \partial_\mu A_\nu(x) + ig[A_\mu(x), A_\nu(x)], \qquad (1.4.26 \ \& \ 4.21)$$

describing the Sagnac interferometer with *platform in rotation*. Unlike the case of the unrotated platform, the rotated platform is Yang–Mills-charged.

In summary, a comparison can be made (Table 4.1) between descriptions of the Sagnac interferometer with platform at rest and with platform in motion.

Turning now to other algebraic approaches to field description: the mathematical algebra offered by *twistors* sheds light on the structure of energy–momentum/angular-momentum of systems.[14,26,27,45] A twistor is a conformally invariant structure, and in Minkowski

---

[aa]$\varphi_{AB} = \varphi_{(AB)}$ (Ref. 27, p. 32).

Table 4.1.

Platform at rest	↔	Platform in motion
Minkowski space–time locally and globally	↔	Minkowski space–time only locally; Minkowski vector spaces as the tangent spaces of space–time events.
Self-dual and anti-self-dual Weyl space–times are complex conjugates	↔	Weyl anti-self-dual space–time independent of self-dual space–time
Conformally flat space–time	↔	Conformally curved space–time
Abelian Maxwell equations apply locally and globally	↔	Abelian Maxwell equations apply locally and Non-Abelian Maxwell equations apply globally
Absence of Weyl tensor	↔	Presence of Weyl tensor
Real space–time	↔	Complex space–time
Twistor space algebra applies	↔	Curved twistor space algebra applies
Fields of SO(3) gauge	↔	Fields of $SU(2)/Z_2$ gauge

space–time it is a pair,

$$Z = (\omega^A, \pi_{A'}), \qquad (4.22)$$

consisting of a spinor field, ω^A, and a complex conjugate spinor field, $\pi_{A'}$, satisfying the twistor equation:

$$\partial_{AA'}\omega^B = -i\varepsilon_A^B \pi_{A'}. \qquad (4.23)$$

However, the extrapolation of twistor algebra to non-Minkowski space–time (as required for a description of the Sagnac interferometer platform in motion) is beyond the scope of the present book. Such a program is addressed by the *ambitwistor* program (cf. the Ward construction in Ref. 27, p. 164–168, and Ref. 45, chap. 9) and covers the topics of curved anti-self-dual space–times, curved twistor spaces and complex space–times.

5. Discussion

We have argued that Coriolis acceleration results in an unmeasurable change in the gauge potentials and a measurable change in the phase factor. Since the phase factor is defined by the gauge potentials, this implies a new definition of the gauge concept. Originally introduced by Weyl as referencing a change in length, the concept was later adapted to quantum–mechanical requirements and henceforth referenced a change in phase. In a further extension, the present usage implies a causal relationship between Coriolis acceleration and the gauge potentials. That being so, there seems to be no reason to discriminate between the *effects* of the electromagnetic **A** potential field on a test particle (such as in the Aharonov–Bohm effect) and the *effects* of acceleration (kinematic, linear or gravitational) on test fields in an interferometer. In all cases, a measurable change in a phase factor results. Because (1) the only measurable indication of the presence of the forces in these cases is indicated by the change in the phase factor of test particles/waves in a global interferometric situation, and (2) we have attributed this phase factor change to a conservation of action law and topological group constraints, there seems reason enough to conclude that the electromagnetic test particle/wave does not, and cannot, discriminate between the different causal (force) origins of a change in the phase factor in (1) and (2). That is, and considering diverse causal origins of the qualitatively identical result, if (a) some change in the space–time metric, or (b) the nearby presence of a mass (gravitational attraction), or (c) linear accelation, or (d) kinematic acceleration, results in the *same* qualitative effect, namely electromagnetic phase factor changes, then there is no need to search for the *unification* of e.g. gravitational and electromagnetic *forces*. Rather, what should be appreciated is that if electromagnetic force fields and the space–time metric gauge fields are bound by conservation of action and topological group constraints, the electromagnetic group, of whatever symmetry, can be perturbed in a number of ways of which (a), (b), (c) and (d) are examples. Furthermore, the electromagnetic gauge group does not discriminate between (a), (b), (c) and (d), for the result of all of these ways of perturbing the group is the qualitatively identical change — a change in the phase factor. It is inappropriate, then, to require unification of electromagnetic force with the "force" of gravity in that gravity is one manifestation of the electromagnetic gauge field registering a perturbation of the force field under condition (b), both force and gauge fields being confined by conservation of action bounded by topological degrees of freedom. Bound by global conformal invariance or conservation of action, *any change* in the conformal structure of *e.m. fields* results in a space–time metric change; but, *pari passu, any change* in the *space–time metric* results in a change in the conformal structure of e.m. fields. Therefore, rather than seeking to unite the *forces* of gravity and electromagnetism, the electromagnetic *force* fields and *gauge* fields (gravity) are already united owing to preservation of action and topological degrees of freedom.

The viewpoint presented here is in conformity with the general theory of relativity. It is also in conformity with the special theory of relativity, *when the platform is at rest, but not when the platform is accelerated or subjected to a Coriolis force, as it should be if the epithet "special" is justified*. The viewpoint presented also implies that the source of the metric fields (of space–time) is the electromagnetic fields and the source of the electromagnetic fields is the metric fields, both being subject to conditions of global conformal invariance.

The present approach to the Sagnac effect, a coherent field effect, is in the spirit of Rainich–Wheeler–Misner's "already unified field theory," considering that RWM adopts the stance that space–time is not just an arena for the electromagnetic fields to play out their dynamic interactions, but both space–time and electromagnetism, together, form a dynamic interactive entity.[22,36,37,38,51,52] However, the present view has taken advantage of the major conceptual advances made in both physics and mathematics since RWM was first proposed. For example, advances have occurred in application of gauge theory and specifically, Yang–Mills theory, with the discovery of the instanton concept, as well as with the recognition that the underlying algebraic logic of fields, as defined by group theory, together with the contributions of Cartan, Weyl and Lie, prescribes those fields' dynamic behavior. We are now able, therefore, to distinguish between gauge potentials and phase factors and even between the gauge potentials a_μ and $\mathbf{A}_\mu$ and the phase factors Φ and Φ^*, and, moreover, to ascribe physical meaning to these distinctions. Drawing on group theory concepts, the present approach does not seek gravitational unity with the *conventional* U(1) formulation of Maxwell's theory, but with the *"compactified"* SU(2) and higher order versions. And again, whereas for RWM the concept of *force* remains central, here *the gauge field* is at least as important. But the major difference is that whereas RWM does not use the electromagnetic potential, the use of this potential is pivotal to the present approach. Nonetheless, despite these fundamental differences and other incompatibilities, the strategic theoretical orientation toward fields, their metric, and the dynamic interaction of the two, is in a similar spirit.

Taken to its logical conclusion, the approach adopted here requires that under the special topological conditions described above, and only under those conditions, the photon associated with the Φ^* field should acquire mass and propagate as a disturbance of the gravitational metric. Using field conversion, a Φ^* field-based mechanism would efficiently propagate energy as well as communications, and penetrate media normally impenetrable to force field photons.

References

1. Aronowitz, F., "The laser gyro," in M. Ross (ed.), *Laser Applications*, Vol. 1 (Academic, New York, 1971).

2. Barrett, T.W., "Comments on the Harmuth Ansatz: Use of a magnetic current density in the calculation of the propagation velocity of signals by amended Maxwell theory," *IEEE Trans. Electromagn. Compat.* Vol. 30, pp. 419–420, 1988.

3. Barrett, T.W., "Electromagnetic phenomena not explained by Maxwell's equations," in A. Lakhtakia (ed.) *Essays on the Formal Aspects of Electromagnetic Theory* (World Scientific, Singapore, 1993), pp. 6–86.

4. Barrett, T.W, "Sagnac effect: A consequence of conservation of action due to gauge field global conformal invariance in a multiply-joined topology of coherent fields," in Barrett, T.W. & Grimes, D.M. (eds.) *Advanced Electromagnetism: Foundations and Applications* (World Scientific, Singapore, 1995), pp. 278–279.

5. Brauer, R. & Weyl, H., "Spinors in n dimensions," *Am. J. Math.*, Vol. 57, pp. 425–449, 1935.

6. Cartan, E., "Les groupes projectifs qui ne laissent invariante aucune multiplicité plane," *Bull. Soc. Math. de France*, Vol. 41, pp. 53–96, 1913.

7. Cartan, E., *The Theory of Spinors* (Dover, New York, 1981) [*Leçons sur la Théorie des Spineurs*, 2 Vols. (Hermann, Paris, 1966)].

8. Chow, W.W., Gea-Banacloche, J., Pedrotti, L.M., Sanders, V.E., Schleich, W. & Scully, M.O., "The ring laser gyro," *Rev. Mod. Phys.*, Vol. 57, pp. 61–104, 1985.

9. de Rham, G., *Variétés Differentiable* (Hermann, Paris, 1960).

10. Forder, P.W., "Ring gyroscopes: An application of adiabatic invariance," *J. Phys. A: Math. Gen.*, Vol. 17, pp. 1343–1355, 1984.

11. Gyorffi, G.L. & Lamb, W.E., *Theory of a Ring Laser* (Michigan Microfilm, Yale University, 1965).

12. Heaviside, O., *Electromagnetic Theory*, 1893 (Chelsea, New York, 1971).

13. Hong-Mo, C. & Tsun, T.S., *Some Elementary Gauge Theory Concepts* (World Scientific, Singapore, 1993).

14. Huggett, S.A. & Tod, K.P., *An Introduction to Twistor Theory* (Cambridge University Press, 1985).

15. Kenyon, I.R., *General Relativity* (Oxford University Press, 1990).

16. Konopinski, E.J., "What the electromagnetic vector potential describes," *Am. J. Phys.*, Vol. 46, pp. 499–502, 1978.

17. Macek, W.M. & Davis, D.T.M., "Rotation rate sensing with traveling wave ring lasers," *Appl. Phys. Lett.*, Vol. 2, pp. 67–68, 1963.

18. Menegozzi, L.M. & Lamb, W.E., "Theory of a ring laser," *Phys. Rev.*, Vol. A8, pp. 2103–2125, 1973.

19. Metz, A., "The problems relating to rotation in the theory of relativity, *J. Phys. Radium,* Vol. 13, pp. 224–238, 1952.

20. Metz, A., "Théorie relativiste de l'expérience de Sagnac avec interposition de tubes réfringents immobiles," *Compt Rend.*, Vol. 234, pp. 597–599, 1952.

21. Metz, A., "Théorie relativiste d'une expérience de Dufour et Prunier," *Compt Rend.*, Vol. 234, pp. 705–707, 1952.

22. Misner, C.W. & Wheeler, J.A., "Classical physics as geometry: Gravitation, electromagnetism, unquantized charge, and mass as properties of curved empty space," *Ann. Phys.*, Vol. 2, pp. 525–603, 1957.

23. Moyer, H.G., "An action principle combining electromagnetism and general relativity," *J. Math. Phys.* Vol. 28, pp. 705–710, 1987.

24. Nash, C & Shen, S., *Topology and Geometry for Physicists* (Academic, New York, 1983).

25. Pegram, G.B., "Unipolar induction and electron theory," *Phys. Rev.*, Vol. 10, pp. 591–600, 1917.

26. Penrose, R. & Rindler, W., *Spinors and Space–Time*, Vol. 1 (Cambridge University Press, 1984).

27. Penrose, R. & Rindler, W., *Spinors and Space–Time*, Vol. 2 (Cambridge University Press, 1986).

28. Plebanski, J., "Electromagnetic waves in gravitational fields," *Phys. Rev.*, Vol. 118, pp. 1396–1408, 1960.

29. Post, E.J., *Formal Structure of Electromagnetics: General Covariance and Electromagnetics* (North-Holland, Amsterdam; John Wiley & Sons, New York, 1962).

30. Post, E.J., "Sagnac effect," *Rev. Mod. Phys.*, Vol. 39, pp. 475–493, 1967.

31. Post, E.J., "Geometry and physics: A global approach," in: M. Bunge (ed.) *Problems in the Foundations of Physics* (Springer-Verlag, New York, 1971), pp. 57–78.

32. Post, E.J., "Interferometric path-length changes due to motion," *J. Opt. Soc. Am.*, Vol. 62, pp. 234–239, 1972a.

33. Post, E.J., "The constitutive map and some of its ramifications," *Ann. Phys.* Vol. 71, pp. 497–518, 1972b.

34. Post, E.J., "Ramifications of flux quantization," *Phys. Rev.*, Vol. D9, pp. 3379–3385, 1974.

35. Post, E.J., "Can microphysical structure be probed by period integrals?," *Phys. Rev.*, Vol. D25, pp. 3223–3229, 1982.

36. Power, E.A. & Wheeler, J.A., "Thermal geons," *Rev. Mod. Phys.*, Vol. 29, pp 480–495, 1957.

37. Rainich, G.Y., "Electrodynamics in the general relativity theory," *Proc. Nat. Acad. Sci.*, Vol. 10, pp. 124–127, 1924.

38. Rainich, G.Y., "Electrodynamics in the general relativity theory," *Trans. Am. Math. Soc.*, Vol. 27, pp. 106–136, 1925.

39. Rosenthal, A.H., "Regenerative circulatory multiple-beam interferometry for the study of light-propagation effects," *J. Opt. Soc. Am.*, Vol. 52, pp. 1143, 1962.

40. Sagnac, G., "L'éther lumineux demonstré par l'effet du vent relatif d'éther dans un interféromètre en rotation uniforme," *Comptes Rendus Acad. Sci.*, Vol. 157, Séance du 27 Oct., pp. 708–710, 1913a.

41. Sagnac, G., "Sur la preuve de la réalité de l'éther lumineaux par l'expérience de l'interférographe tournant," *Comptes Rendus Acad. Sci.*, Séance du 22 Déc., Vol. 157, pp. 1410–1413, 1913b.

42. Sagnac, G., "Circulation of the ether in rotating interferometer," *Le Journal de Physique et Le Radium,* 5th Ser. 4, pp. 177–195, 1914.

43. von Laue, M., "Velocity of light in moving bodies," *Ann. Phys.*, Vol. 62, 448–463, 1920.

44. Wald, R.M., *General Relativity* (University of Chicago Press, 1984).
45. Ward, R.S. & Wells, R.O., *Twistor Geometry and Field Theory* (Cambridge University Press, 1990).
46. Weyl, H., "Gravitation und electrizität," *Sitz. Ber. Preuss. Ak. Wiss.*, 465–480, 1918a.
47. Weyl, H., "Reine infinitesimalgeometrie," *Math. Zeit.*, Vol. 2, pp. 384–411, 1918b.
48. Weyl, H., "Gravitation and the electron," *Proc. Nat. Acad. Sci. USA*, Vol. 15, pp. 323–334, 1929.
49. Weyl, H., *The Classical Groups, their Invariants and Representations* (Princeton University Press, 1939, 1946).
50. Weyl, H., *Gesammelte Abhandlungen*, 4 Vols., K. Chandrasekharan (ed.) (Springer-Verlag, Heidelberg, 1968).
51. Wheeler, J.A., "Geons," *Phys. Rev.*, Vol. 97, pp. 511–536, 1955.
52. Wheeler, J.A., *Geometrodynamics* (Academic Press, 1962).
53. Whittaker, E.T., *A History of the Theories of Aether and Electricity*, 2 Vols., 1951 and 1953 (Dover, 1989).
54. Wu, T.T. & Yang, C.N., "Concept of nonintegrable phase factors and global formulation of gauge fields," *Phys. Rev.*, Vol. D12, pp. 3845, 1975.
55. Yang, C.N. & Mills, R.N., "Isotopic spin conservation and generalized gauge invariance," *Phys. Rev.*, Vol. 95, M7, pp. 631, 1954a.
56. Yang, C.N. & Mills, R.N., "Conservation of isotopic spin and isotopic gauge invariance." *Phys. Rev.*, Vol. 96, pp. 191–195, 1954b.
57. Zernike, F., "The convection of light under various circumstances with special reference to Zieman's experiments," *Physica*, Vol. 13, pp. 279–288, 1947.

3

Topological Approaches to Electromagnetism[bb]

Overview

Topology addresses those properties, often associated with *invariant qualities*, which are not altered by continuous deformations. Objects are topologically equivalent, or *homomorphic*, if one object can be changed into another by bending, stretching, twisting, or any other continuous deformation or mapping. Continuous deformations are allowed, but prohibited are foldings which bring formerly distant points into direct contact or overlap, and cutting — unless followed by a regluing, reestablishing the preexisting relationships of continuity.

   The continuous deformations of topology are commonly described in differential equation form and the quantities conserved under the transformations commonly described by differential equations exemplifying an algebra describing operations which preserve that algebra. Evariste Galois (1811–1832) first gave the criteria that an algebraic

---

[bb]Based on: Barrett, T.W., "Topological approaches to electromagnetism," in *Modern Nonlinear Optics, Part 3*, 2nd edn. (Wiley, 2001), pp. 669–734.

equation must satisfy in order to be solvable by radicals. This branch of mathematics came to be known as Galois or *group theory*.

Beginning with G.W. Leibniz in the 17th century, L. Euler in the 18th, B. Reimann, J.B. Listing and A.F. Möbius in the 19th and H. Poincaré in the 20th, *"analysis situs"* (Riemann) or *"topology"* (Listing[cc]) has been used to provide answers to questions concerning what is most fundamental in physical explanation. That question itself implies the question concerning what mathematical structures one uses with confidence to adequately "paint" or describe physical models built from empirical facts. For example, differential equations of motion cannot be fundamental, because they are dependent on boundary conditions which must be justified — usually by group-theoretical considerations. Perhaps, then, group theory[dd] is fundamental.

Group theory certainly offers an austere shorthand for fundamental transformation rules. But it appears to the present writer that the final judge of whether a mathematical group structure can, or cannot, be applied to a physical situation is the topology of that physical situation. Topology dictates and justifies the group transformations.

So, for the present writer, the answer to the question of what is the most fundamental physical description is that it is a description of the topology of the situation. With the topology known, the group theory description is justified and equations of motion can then be justified and defined in specific differential equation form. If there is a requirement for an understanding more basic than the topology of the situation, then all that is left is verbal description of visual images. So we commence an examination of electromagnetism under the assumption that topology defines group transformations and the group transformation rules justify the algebra underlying the differential equations of motion.

Differential equations or a set of differential equations describe a *system* and its evolution. Group symmetry principles summarize

---

[cc]Johann Benedict Listing (1808–1882). See his *Vorstudien zur Topologie* (1847), where, for the first time, the title "Topology" (in German) appeared in print.

[dd]Here we address the kind of groups addressed in Yang–Mills theory, which are *continuous* groups (as opposed to *discrete* groups). Unlike discrete groups, continuous groups contain an infinite number of elements and can be differentiable or analytic. Cf. Yang, C.N. & Mills, R.L., "Conservation of isotopic spin and isotopic gauge invariance," *Phys. Rev.*, vol. 96, pp. 191–195, 1954.

both invariances and the laws of nature independent of a system's specific dynamics. It is necessary that the symmetry transformations be continuous or specified by a set of parameters which can be varied continuously. The symmetry of continuous transformations leads to conservation laws.

There are a variety of special methods used to solve ordinary differential equations. It was Sophus Lie (1842–1899) who showed that all the methods are special cases of integration procedures which are based on the invariance of a differential equation under a continuous group of symmetries. These groups became known as Lie groups.[ee] A symmetry group of a system of differential equations is a group which transforms solutions of the system to other solutions.[ff] In other words, there is an invariance of a differential equation under a transformation of independent and dependent variables. This invariance results in a diffeomorphism on the space of independent and dependent variables, permitting the mapping of solutions to solutions.[gg]

The relationship was made more explicit by Emmy (Amalie) Noether (1882–1935) in theorems now known as *Noether's theorems*,[hh] which related symmetry groups of a variational integral to properties of its associated Euler–Lagrange equations. The most important consequences of this relationship are that (i) conservation of energy arises from invariance under a group of time translations; (ii) conservation of linear momentum arises from invariance under (spatial) translational groups; (iii) conservation of angular momentum arises from invariance under (spatial) rotational groups; and

---

[ee]**Lie Group Algebras**

If a topological group is a group and also a topological space in which group operations are continuous, then Lie groups are topological groups which are also analytic manifolds on which the group operations are analytic.

In the case of Lie algebras, the parameters of a product are analytic functions of the parameters of each factor in the product. For example, $L(\gamma) = L(\alpha)L(\beta)$, where $\gamma = f(\alpha, \beta)$. This guarantees that the group is differentiable. The Lie groups used in Yang–Mills theory are *compact* groups, i.e. the parameters range over a closed interval.

[ff]Cf. Olver, P.J., *Applications of Lie Groups to Differential Equations* (Springer Verlag, 1986).
[gg]Baumann, G., *Symmetry Analyis of Differential Equations with Mathematica* (Springer Verlag, 1998).

[hh]Noether, E., "Invariante variations probleme," *Nachr. Ges. Wiss. Goettingen, Math.-Phys.*, Kl. 171, pp. 235–257, 1918.

(iv) conservation of charge arises from invariance under change of phase of the wave function of charged particles. Conservation and group symmetry laws have been vastly extended to other systems of equations, such as the standard model of modern high energy physics, and also, of importance to the present interest, soliton equations. For example, the *Korteweg–de Vries* "soliton" equation[ii] yields a symmetry algebra spanned by the four vector fields of (i) space translation, (ii) time translation, (iii) Galilean translation, and (iv) scaling.

An aim of the present book is to show that the space-time topology defines electromagnetic field equations[jj] — whether the fields are of force or of phase. That is to say, the premise of this enterprise is that a set of field equations are only valid with respect to a set defined topological description of the physical situation. In particular, the writer has addressed this demonstrating that the A_μ potentials, $\mu = 0, 1, 2, 3$, are not just a mathematical convenience, but — *in certain well-defined situations* — are measurable, i.e. physical. Those situations in which the A_μ potentials are measurable possess a topology the transformation rules of which are describable by the SU(2) group[kk] or higher order groups; and those situations in which the A_μ potentials are not

[ii]Korteweg, D.J. & de Vries, G, "On the change of form of long waves advancing in a rectangular canal, and on a new type of long stationary wave." *Philos. Mag.*, vol. 39, pp. 422–443, 1895.

[jj]Barrett, T.W., "Maxwell's theory extended. Part I: Empirical reasons for questioning the completeness of Maxwell's theory — effects demonstrating the physical significance of the *A* potentials," *Annales de la Fondation Louis de Broglie*, vol. 15, pp. 143–183, 1990;

———, "Maxwell's theory extended. Part II. Theoretical and pragmatic reasons for questioning the completeness of Maxwell's theory," *Annales de la Fondation Louis de Broglie*, vol. 12, pp. 253–283, 1990;

———, "The Ehrenhaft–Mikhailov effect described as the behavior of a low energy density magnetic monopole instanton," *Annales de la Fondation Louis de Broglie*, vol. 19, pp. 291–301, 1994;

———, "Electromagnetic phenomena not explained by Maxwell's equations," in: Lakhtakia, A. (ed.), *Essays on the Formal Aspects of Maxwell's Theory* (World Scientific, Singapore, 1993), pp. 6–86;

———, "Sagnac effect," in: Barrett, T.W. & Grimes, D.M. (eds.), *Advanced Electromagnetism: Foundations, Theory, Applications* (World Scientific, Singapore, 1995), pp. 278–313;

———, "The toroid antenna as a conditioner of electromagnetic fields into (low energy) gauge fields," in *Speculations in Science and Technology*, vol. 21, no.4, pp. 291–320, 1998.

[kk]**SU(*n*) Group Algebra**

Unitary transformations, U(*n*), leave the modulus squared of a complex wave function invariant. The elements of a U(*n*) group are represented by $n \times n$ unitary matrices with a determinant equal to ±1. Special unitary matrices are elements of unitary matrices which leave the determinant equal to +1. There are $n^2 - 1$ independent parameters. SU(n) is a subgroup of U(n) for which the determinant equals +1.

SL(2,C) Group Algebra

The special linear group of 2 × 2 matrices of determinant 1 with complex entries is SL(2,C).

SU(2) Group Algebra

SU(2) is a subgroup of SL(2,C). The are $2^2 - 1 = 3$ independent parameters for the special unitary group SU(2) of 2 × 2 matrices. SU(2) is a Lie algebra such that for the angular momentum generators, J_i, the commutation relations are $[J_i, J_j] = i\varepsilon_{ijk}J_k$, $i, j, k = 1, 2, 3$. The SU(2) group describes rotation in three-dimensional space with two parameters (see below). There is a well-known SU(2) matrix relating the Euler angles of O(3) and the complex parameters of SU(2):

$$\cos\left[\tfrac{\beta}{2}\right]\exp\left[\tfrac{i(\alpha+\gamma)}{2}\right] \quad \sin\left[\tfrac{\beta}{2}\right]\exp\left[\tfrac{-(\alpha-\gamma)}{2}\right]$$
$$-\sin\left[\tfrac{\beta}{2}\right]\exp\left[\tfrac{i(\alpha-\gamma)}{2}\right] \quad \cos\left[\tfrac{\beta}{2}\right]\exp\left[\tfrac{-i(\alpha+\gamma)}{2}\right],$$

where α, β, γ are the Euler angles. It is also well known that a homomorphism exists between O(3) and SU(2), and the elements of SU(2) can be associated with rotations in O(3); and SU(2) is the *covering group* of O(3). Therefore, it is easy to show that SU(2) can be obtained from O(3). These SU(2) transformations define the relations between the Euler angles of group O(3) with the parameters of SU(2). For comparison with the above, if the rotation matrix $R(\alpha, \beta, \gamma)$ in O(3) is represented as

$$\begin{pmatrix} \cos[\alpha]\cos[\beta]\cos[\gamma]-\sin[\alpha]\sin[\gamma] & \sin[\alpha]\cos[\beta]\cos[\gamma]+\cos[\alpha]\sin[\gamma] & -\sin[\beta]\cos[\gamma] \\ -\cos[\alpha]\cos[\beta]\sin[\gamma]-\sin[\alpha]\cos[\gamma] & -\sin[\alpha]\cos[\beta]\sin[\gamma]+\cos[\alpha]\cos[\gamma] & \sin[\beta]\sin[\gamma] \\ \cos[\alpha]\sin[\beta] & \sin[\alpha]\sin[\beta] & \cos[\beta] \end{pmatrix},$$

then the orthogonal rotations about the coordinate axes are

$$R_1(\alpha) = \begin{pmatrix} \cos[\alpha] & \sin[\alpha] & 0 \\ -\sin[\alpha] & \cos[\alpha] & 0 \\ 0 & 0 & 1 \end{pmatrix} \quad R_2(\beta) = \begin{pmatrix} \cos[\beta] & 0 & -\sin[\beta] \\ 0 & 1 & 0 \\ \sin[\beta] & 0 & \cos[\beta] \end{pmatrix}$$

$$R_3(\gamma) = \begin{pmatrix} \cos[\gamma] & \sin[\gamma] & 0 \\ -\sin[\gamma] & \cos[\gamma] & 0 \\ 0 & 0 & 1 \end{pmatrix}.$$

An isotropic parameter, ϖ, can be defined as

$$\varpi = \frac{x - iy}{z},$$

where x, y, z are the spatial coordinates. If ϖ is written as the quotient of μ_1 and μ_2, or the homogeneous coordinates of the bilinear transformation, then

$$|\mu_1'\mu_2'\rangle = \begin{bmatrix} \cos\left[\tfrac{\beta}{2}\right]\exp\left[\tfrac{i(\alpha+\gamma)}{2}\right] & \sin\left[\tfrac{\beta}{2}\right]\exp\left[\tfrac{-(\alpha-\gamma)}{2}\right] \\ -\sin\left[\tfrac{\beta}{2}\right]\exp\left[\tfrac{i(\alpha-\gamma)}{2}\right] & \cos\left[\tfrac{\beta}{2}\right]\exp\left[\tfrac{-i(\alpha+\gamma)}{2}\right] \end{bmatrix} |\mu_1\mu_2\rangle,$$

which is the relation between the Euler angles of O(3) and the complex parameters of SU(2). However, there is not a unique one-to-one relation, for two rotations in O(3) correspond to one direction in SU(2). There is thus a many-to-one or homomorphism between O(3) and SU(2).

In the case of a complex two-dimensional vector (u, v),

$$\begin{pmatrix} u' \\ v' \end{pmatrix} = \begin{pmatrix} \cos\left[\frac{\beta}{2}\right]\exp\left[\frac{i(\alpha+\gamma)}{2}\right] & \sin\left[\frac{\beta}{2}\right]\exp\left[\frac{-(\alpha-\gamma)}{2}\right] \\ -\sin\left[\frac{\beta}{2}\right]\exp\left[\frac{i(\alpha-\gamma)}{2}\right] & \cos\left[\frac{\beta}{2}\right]\exp\left[\frac{-i(\alpha+\gamma)}{2}\right] \end{pmatrix} \begin{pmatrix} u \\ v \end{pmatrix}.$$

If we define

$$a = \cos\left[\frac{\beta}{2}\right]\exp\left[\frac{i(\alpha+\gamma)}{2}\right],$$

$$b = \sin\left[\frac{\beta}{2}\right]\exp\left[\frac{-(\alpha-\gamma)}{2}\right],$$

then

$$|\mu_1'\mu_2'\rangle = \begin{bmatrix} a & b \\ -b^* & a^* \end{bmatrix} |\mu_1\mu_2\rangle,$$

where

$$\begin{bmatrix} a & b \\ -b^* & a^* \end{bmatrix}$$

are the well-known SU(2) transformation rules. Defining $c = -b^*$ and $d = a^*$, we have the determinant

$$ad - bc = 1 \quad \text{or} \quad aa^* - b(-b^*) = 1.$$

Defining the (x, y, z) coordinates with respect to a complex two-dimensioanl vector (u, v) as

$$x = \frac{1}{2}(u^2 - v^2), \quad y = \frac{1}{2i}(u^2 + v^2), \quad z = uv,$$

SU(2) transformations leave the squared distance $x^2 + y^2 + z^2$ invariant.

Every element of SU(2) can be written as:

$$\begin{bmatrix} a & b \\ -b^* & a^* \end{bmatrix}, \quad |a|^2 + |b|^2 = 1.$$

Defining

$$a = y_1 - iy_2, \quad b = y_3 - iy_4,$$

the parameters y_1, y_2, y_3, y_4 indicate positions in SU(2) with the constraint

$$y_1^2 + y_2^2 + y_3^2 + y_4^2 = 1,$$

which indicates that the group SU(2) is a three-dimensional unit sphere in the four-dimensional y space. This means that any closed curve on that sphere can be shrunk to a point. In other words, SU(2) is *simply connected*.

It is important to note that SU(2) is the quantum-mechanical "rotation group."

Homomorphism of O(3) and SU(2)

There is an important relationship between O(3) and SU(2). The elements of SU(2) are associated with rotations in three-dimensional space. To make this relationship explicit, new coordinates are defined

$$x = \frac{1}{2}(u^2 - v^2); \quad y = \frac{1}{2i}(u^2 + v^2); \quad z = uv.$$

Explicitly, the SU(2) transformations leave the squared three-dimensional distance $x^2 + y^2 + z^2$ invariant, and invariance which relates three-dimensional rotations to elements of SU(2). If a, b

measurable possess a topology the transformation rules of which are describable by the U(1) group.[ll]

---

of the elements of SU(2) are defined,

$$a = \cos\frac{\beta}{2}\exp\frac{i\,(\alpha + \gamma)}{2}, \quad b = \sin\frac{\beta}{2}\exp\frac{-i\,(\alpha - \gamma)}{2},$$

then the general rotation matrix $R(\alpha, \beta, \gamma)$ can be associated with the SU(2) matrix,

$$\begin{pmatrix} \cos\frac{\beta}{2}\exp\frac{i(\alpha+\gamma)}{2} & \sin\frac{\beta}{2}\exp\frac{-i(\alpha-\gamma)}{2} \\ -\sin\frac{\beta}{2}\exp\frac{i(\alpha-\gamma)}{2} & \cos\frac{\beta}{2}\exp\frac{-i(\alpha+\gamma)}{2} \end{pmatrix},$$

by means of the Euler angles.

It is important to indicate that this matrix does not give a unique one-to-one relationship between the general rotation matrix $R(\alpha, \beta, \gamma)$ and the SU(2) group. This can be seen if (i) we let $\alpha = 0, \beta = 0, \gamma = 0$, which gives the matrix

$$\begin{pmatrix} 1 & 0 \\ 0 & 1 \end{pmatrix},$$

and (ii) $\alpha = 0, \beta = 2\pi, \gamma = 0$, which gives the matrix

$$\begin{pmatrix} -1 & 0 \\ 0 & -1 \end{pmatrix}.$$

Both matrices define zero rotation in three-dimensional space, so we see that this zero rotation in three-dimensional space corresponds to two different SU(2) elements depending on the value of β. There is thus a *homomorphism*, or many-to-one mapping relationship, between O(3) and SU(2) — where "many" is two in this case — but not a one-to-one mapping.

SO(2) Group Algebra

The collection of matrices in Euclidean two-dimensional space (the plane) which are orthogonal and for which the determinant is +1 is a subgroup of O(2). SO(2) is the special orthogonal group in two variables.

The rotation in the plane is represented by the SO(2) group

$$R(\theta) = \begin{pmatrix} \cos[\theta] & -\sin[\theta] \\ \sin[\theta] & \cos[\theta] \end{pmatrix} \quad \theta \in \Re,$$

where $R(\theta)R(\gamma) = R(\theta + \gamma)$. S^1, or the unit circle in the complex plane with multiplication as the group operation is an SO(2) group.

[ll]U(n) Group Algebra

Unitary matrices, U, have a determinant equal to ± 1. The elements of U(n) are represented by $n \times n$ unitary matrices.

U(1) Group Algebra

The one-dimensional unitary group, or U(1), is characterized by one continuous parameter. U(1) is also differentiable and the derivative is also an element of U(1). A well-known example of a U(1) group is that of all the possible phases of a wave function, which are angular coordinates in a two-dimensional space. When interpreted in this way — as the internal phase of the U(1) group of electromagnetism — the U(1) group is merely a circle $(0 - 2\pi)$.

Historically, electromagnetic theory was developed for situations described by the U(1) group. The dynamic equations describing the transformations and interrelationships of the force field are the well known Maxwell equations, and the group algebra underlying these equations is U(1). There was a need to extend these equations to describe SU(2) situations and to derive equations whose underlying algebra is SU(2). These two formulations were provided in previous chapters and are shown again here in Table 1. Table 2 shows the electric charge density, ρ_e, the magnetic charge density, ρ_m, the electric current density, g_e, the magnetic current density, g_m, the electric conductivity, σ, and the magnetic conductivity, s.

In the following sections, four topics are addressed: the mathematical entities, or waves, called *solitons*; the mathematical entities called *instantons*; a beam — an electromagnetic wave — which is *polarization-modulated over a set sampling interval*; and the *Aharonov–Bohm effect*. Our intention is to show that these entities, waves and effects can only be adequately characterized and differentiated, and thus understood, by using topological characterizations. Once they are characterized, the way becomes open for control or engineering of these entities, waves and effects.

Table 1.   Maxwell Equations in U(1) and SU(2) Symmetry Forms

	U(1) Symmetry Form (Traditional Maxwell Equations)	SU(2) Symmetry Form
Gauss's law	$\nabla \cdot E = J_0$	$\nabla \cdot E = J_0 - iq(A \cdot E - E \cdot A)$
Ampère's law	$\frac{\partial E}{\partial t} - \nabla \times B + J = 0$	$\frac{\partial E}{\partial t} - \nabla \times B + J + iq[A_0, E]$ $-iq(A \times B - B \times A) = 0$
Coulomb's law	$\nabla \cdot B = 0$	$\nabla \cdot B + iq(A \cdot B - B \cdot A) = 0$
Faraday's law	$\nabla \times E + \frac{\partial B}{\partial t} = 0$	$\nabla \times E + \frac{\partial B}{\partial t} + iq[A_0, B]$ $+iq(A \times E - E \times A) = 0$

Table 2. The U(1) and SU(2) Symmetry Forms of the Major Variables

U(1) Symmetry Form (Traditional Maxwell Theory)	SU(2) Symmetry Form
$\rho_e = J_0$	$\rho_e = J_0 - iq(A \cdot E - E \cdot A) = J_0 + qJ_z$
$\rho_m = 0$	$\rho_m = -iq(A \cdot B - B \cdot A) = -iqJ_y$
$g_e = J$	$g_e = iq[A_0, E] - iq(A \times B - B \times A) + J$
	$\quad = iq[A_0, E] - iqJ_x + J$
$g_m = 0$	$g_m = iq[A_0, B] - iq(A \times E - E \times A) = iq[A_0, B] - iqJ_z$
$\sigma = J/E$	$\sigma = \frac{\{iq[A_0, E] - iq(A \times B - B \times A) + J\}}{E} = \frac{\{iq[A_0, E] - iqJ_x + J\}}{E}$
$s = 0$	$s = \frac{\{iq[A_0, B] - iq(A \times E - E \times A)\}}{H} = \frac{\{iq[A_0, B] - iqJ_z\}}{H}$

1. Solitons[mm]

Soliton solutions to differential equations require complete integrability and integrable systems to conserve geometric features related to symmetry. Unlike the equations of motion for conventional Maxwell theory, which are solutions of U(1) symmetry systems, solitons are solutions of SU(2) symmetry systems. These notions of group symmetry are more fundamental than differential equation descriptions. Therefore, although a complete exposition is beyond the scope of the present review, we develop some basic concepts in order to place differential equation descriptions within the context of group theory.

Within this context, *ordinary differential equations are viewed as vector fields on manifolds or configuration spaces.*[nn]. For example, Newton's equations are second order differential equations describing smooth curves on Riemannian manifolds. Noether's theorem[oo] states

---

[mm]A soliton is a solitary wave which preserves its shape and speed in a collision with another solitary wave. Cf. Barrett, T.W., in: Taylor, J.D. (ed.), *Introduction to Ultra-Wideband Radar Systems* (CRC, Boca Raton, 1995), pp. 404–413: Infeld, E. & Rowlands, G., *Nonlinear Waves, Solitons and Chaos*, 2nd edn. (Cambridge University Press, 2000).

[nn]Cf. Olver, P.J., *Applications of Lie Groups to Differential Equations* (Springer Verlag, 1986).

[oo]Noether, E., "Invariante Variations Probleme," *Nachr. Ges. Wiss. Goettingen, Math.-Phys. Kl.* 171, pp. 235–257, 1918.

that a diffeomorphism,[pp] ϕ, of a Riemannian manifold, C, indicates a diffeomorphism, $D\phi$, of its tangent[qq] bundle,[rr] TC. If ϕ is a symmetry of Newton's equations, then $D\phi$ preserves the Lagrangian, i.e.

$$\mathcal{L} \circ \mathcal{D}\phi = \mathcal{L}.$$

As opposed to equations of motion in conventional Maxwell theory, *soliton flows are Hamiltonian flows*. Such Hamiltonian functions define *symplectic structures*[ss] for which there is an absence of *local* invariants but an infinite-dimensional group of diffeomorphisms which preserve *global* properties. In the case of solitons, the global properties are those permitting the matching of the nonlinear and dispersive characteristics of the medium through which the wave moves.

In order to achieve this match, two linear operators, L and A, are postulated to be associated with a partial differential equation (PDE). The two linear operators are known as the *Lax pair*. The operator L is defined by

$$L = \frac{\partial^2}{\partial x^2} + u(x, t),$$

with a related eigenproblem:

$$L\psi + \lambda\psi = 0. \tag{1.1}$$

The temporal evolution of ψ is defined as

$$\psi_t = -A\psi, \tag{1.2}$$

[pp] A *diffeomorphism* is an elementary concept of topology and is important to the understanding of differential equations. It can be defined in the following way:

If the sets U and V are open sets both defined over the space R^m, i.e. $U \subset R^m$ is open and $V \subset R^m$ is open, where "open" means "nonoverlapping," then the mapping $\psi : U \to V$ is an infinitely differentiable map with an infinitely differential inverse, and objects defined in U will have equivalent counterparts in V. The mapping ψ is a diffeomorphism. It is a smooth and infinitely differentiable function. The important point is: conservation rules apply to diffeomorphisms, because of their infinite differentiability. Therefore diffeomorphisms constitute fundamental characterizations of differential equations.

[qq] A vector field on a manifold, M, gives a *tangent vector* at each point of M.

[rr] A *bundle* is a structure consisting of a manifold E, and manifold M, and an onto map: $\pi : E \to M$.

[ss] *Symplectic topology* is the study of the global phenomena of symplectic symmetry. Symplectic symmetry structures have no local invariants. This is a subfield of topology; see e.g. McDuff, D. & Salamon, D., *Introduction to Symplectic Topology* (Clarendon, Oxford, 1995).

with the operator of the form

$$A = a_0 \frac{\partial^n}{\partial x^n} + a_1 \frac{\partial^{n-1}}{\partial x^{n-1}} + \cdots + a_n,$$

where a_0 is a constant and the n coefficients a_i are functions of x and t. Differentiating (1.1) gives

$$L_t \psi + L \psi_t = -\lambda_t \psi - \lambda \psi_t.$$

Inserting (1.2),

$$L \psi_t = -LA \psi$$

or

$$\lambda \psi_t = AL \psi.$$

Using (1.1) again,

$$[L, A] = LA - AL = L_t + \lambda_t, \tag{1.3}$$

and for a time-independent λ,

$$[L, A] = L_t.$$

This equation provides a method for finding A.

Translating the above into a group theory formulation: in order to relate the three major soliton equations to group theory it is necessary to examine the *Lax equation*[tt] (1.3) as a the *zero-curvature condition* (ZCC). The ZCC expresses the flatness of a connection by the

---

[tt]Lax, P.D., "Integrals of nonlinear equations of evolution and solitary waves," *Comm. Pure Appl. Math.*, vol. 21, pp. 467–490, 1968;

Lax, P.D., "Periodic solutions of the KdV equations," in *Nonlinear Wave Motion: Lectures in Applied Math.* (American Mathematical Society, 1974), vol. 15, pp. 85–96.

commutation relations of the covariant derivative operators[uu] and, in terms of the Lax equation, is

$$L_t - A_x - [L, A] = 0,$$

or[uu]

$$\left[\frac{\partial}{\partial x} - L \frac{\partial}{\partial t} - A \right] = 0,$$

or

$$\left(\frac{\partial}{\partial x} - L \right)_t = \left[A \frac{\partial}{\partial x} - L \right].$$

More recently, Palais[uu] showed that the generic cases of the soliton — the *Korteweg–de Vries equation* (KdV), the *nonlinear Schrödinger equation* (NLS) and the *sine-Gordon equation* (SGE) — can be given an SU(2) formulation. In each of the three cases considered below, V is a one-dimensional space that is embedded in the space of off-diagonal complex matrices, $\left(\begin{smallmatrix} 0 & b \\ c & 0 \end{smallmatrix} \right)$, and in each case $L(u) = a\lambda + u$, where u is a potential, λ is a complex parameter, and a is the constant, diagonal, trace zero matrix,

$$a = \begin{pmatrix} -i & 0 \\ 0 & i \end{pmatrix}.$$

The matrix definition of a links these equations to an SU(2) formulation. (Other matrix definitions of a could, of course, link a to higher group symmetries.)

To carry out this objective, an inverse scattering theory function is defined[tt]:

$B(\xi) = \sum_{n=1}^{N} c_n^2 \exp[-\kappa_n \xi] + \frac{1}{2\pi} \int_{-\infty}^{+\infty} b(k) \exp[ik\xi]dk,$ where
$-\kappa_1^2, \ldots, -\kappa_N^2$ are discrete eigenvalues of u,
$c_1, \ldots, c_N$ are normalizing constants, and
$b(k)$ are reflection coefficients.

[uu]Palais, R.S., "The symmetries of solitons," *Bull. Am. Math. Soc.*, vol. 34, pp. 339–403, 1997.

Therefore, in a *first case* (the KDV), if

$$u(x) = \begin{pmatrix} 0 & q(x) \\ -1 & 0 \end{pmatrix} \quad \text{and}$$

$$B(u) = a\lambda^3 + u\lambda^2 + \begin{pmatrix} \dfrac{i}{2}q & \dfrac{i}{2}q_x \\ 0 & -\dfrac{i}{2}q \end{pmatrix}\lambda + \begin{pmatrix} \dfrac{q_x}{4} & \dfrac{-q^2}{2} \\ \dfrac{q}{2} & \dfrac{-q_x}{4} \end{pmatrix},$$

then the ZCC (Lax equation) is satisfied if and only if q satisfies the KDV in the form $q_t = -\frac{1}{4}(6qq_x + q_{xxx})$.

In a *second case* (the NLS), if

$$u(x) = \begin{pmatrix} 0 & q(x) \\ -\bar{q}(x) & 0 \end{pmatrix} \quad \text{and}$$

$$B(u) = a\lambda^3 + u\lambda^2 + \begin{pmatrix} \dfrac{i}{2}|q|^2 & \dfrac{i}{2}q_x \\ -\dfrac{i}{2}\bar{q}_x & -\dfrac{i}{2}|q|^2 \end{pmatrix},$$

then the ZCC (Lax equation) is satisfied if and only if $q(x, t)$ satisfies the NLS in the form $q_t = \frac{i}{2}(q_{xx} + 2|q|^2 q)$.

In a *third case* (the SGE), if

$$u(x) = \begin{pmatrix} 0 & -\dfrac{q_x(x)}{2} \\ \dfrac{q_x(x)}{2} & 0 \end{pmatrix} \quad \text{and}$$

$$B(u) = \frac{i}{4\lambda}\begin{pmatrix} \cos[q(x)] & \sin[q(x)] \\ \sin[q(x)] & -\cos[q(x)] \end{pmatrix},$$

then the ZCC (Lax equation) is satisfied if and only if q satisfies the SGE in the form $q_t = \sin[q]$.

With the connection of PDEs, and especially soliton forms, to group symmetries established, one can conclude that *if* the Maxwell equation of motion which includes electric *and* magnetic conductivity is in soliton (SGE) form, the group symmetry of the Maxwell field is

SU(2). Furthermore, because solitons define Hamiltonian flows, their energy conservation is due to their *symplectic structure*.

In order to clarify the difference between conventional Maxwell theory, which is of U(1) symmetry, and Maxwell theory extended to SU(2) symmetry, we can describe both in terms of mappings of a field $\psi(x)$. In the case of U(1) Maxwell theory, a mapping $\psi \to \psi'$ is

$$\psi(x) \to \psi'(x) = \exp[ia(x)]\,\psi(x),$$

where $a(x)$ is the conventional vector potential. However, in the case of SU(2) extended Maxwell theory, a mapping $\psi \to \psi'$ is

$$\psi(x) \to \psi'(x) = \exp[iS(x)]\,\psi(x),$$

where $S(x)$ is the action and an element of an SU(2) field defined as

$$S(x) = \int \mathbf{A}dx,$$

where $\mathbf{A}$ is *the matrix form of the vector potential*. Therefore, we see the necessity *to adopt a matrix formulation of the vector potential when addressing* SU(2) *forms of Maxwell theory*.

2. Instantons

Instantons[vv] correspond to the minima of the Euclidean action and are pseudoparticle solutions[ww] of SU(2) Yang–Mills equations in Euclidean four-space.[xx] A complete construction for any Yang–Mills group is also available.[yy] In other words:

> "It is reasonable. . .to ask for the determination of the classical field configurations in Euclidean space which minimize the action, subject to appropriate asymptotic conditions in four-space. These classical solutions are the instantons of the Yang–Mills theory."[zz]

[vv]Cf. Jackiw, R., Nohl, C. & Rebbi, C. *Classical and Semi-classical Solutions to Yang–Mills Theory, Proc.* 1977 Banff School (Plenum).

[ww]Belavin, A., Polyakov, A., Schwartz, A. & Tyupkin, Y., "Pseudoparticle solutions of the Yang–Mills equations," *Phys. Lett.*, vol. 59B, pp. 85–87, 1975.

[xx]Cf. Atiyah, M.F. & Ward, R.S., "Instantons and algebraic geometry," *Commun. Math. Phys.*, vol. 55, pp. 117–124, 1977.

[yy]Atiyah, M.F., Hitchin, N.J., Drinfeld, V.G. & Manin, Yu.I., "Construction of instantons," *Phys. Lett.*, vol. 65A, pp. 23–25, 1978.

[zz]Atiyah, M., in *Michael Atiyah: Collected Works, Volume 5, Gauge Theories* (Clarendon, Oxford, 1988), p. 80.

In the light of the intention of this book to further the use of topology in electromagnetic theory, we quote further:

> "If one were to search ab initio for a non-linear generalization of Maxwell's equation to explain elementary particles, there are various symmetry group properties one would require. These are
>
> (i) *external symmetries* under the Lorentz and Poincaré groups and under the conformal group if one is taking the rest-mass to be zero,
>
> (ii) *internal symmetries* under groups like SU(2) or SU(3) to account for the known features of elementary particles,
>
> (iii) *covariance* or the ability to be coupled to gravitation by working on curved space–time."[aaa]

In this book, the instanton concept in electromagnetism is applied for the following two reasons: (1) in some sense, the instanton, or pseudoparticle, is a compactification of degrees of freedom due to the particle's boundary conditions; and (2) the instanton, or pseudoparticle, then exhibits the behavior (the transformation or symmetry rules) of a high energy particle, *but without the presence of high energy*, i.e. the pseudoparticle shares certain behavioral characteristics in common (shares transformation rules, and hence symmetry rules in common) with a particle of much higher energy.

Therefore, we have suggested[bbb] that the Mikhailov effect,[ccc] and the Ehrenhaft (1879–1952), effect, which address demonstrations exhibiting *magnetic-charge-like behavior*, are examples of instanton or pseudoparticle behavior. Stated differently: (1) the instanton shows that there are ways other than possession of high energy to achieve high symmetry states; and (2) symmetry dictates behavior.

[aaa]Atiyah, M., in *Michael Atiyah: Collected Works, Volume 5, Gauge Theories* (Clarendon, Oxford, 1988).

[bbb]Barrett, T.W., "The Ehrenhaft–Mikhailov effect described as the behavior of a low energy density magnetic monopole instanton." *Annales de la Fondation Louis de Broglie*, vol. 19, pp. 291–301, 1994.

[ccc]A summary of the Mikhailov effect is: in Barrett, T.W. & Grimes, D.M. (eds.), *Advanced Electromagnetism: Foundations, Theory & Applications* (World Scientific, Singapore, 1995), pp. 593–619.

3.   Polarization Modulation Over a Set Sampling Interval[ddd]

It is well known that all static polarizations of a beam of radiation, as well as all static rotations of the axis of that beam, can be represented on a Poincaré sphere[eee] [Fig. 3.1(A)]. A vector can be placed in the middle of the sphere and pointed to the underside of the surface of the sphere at a location on the surface which represents the *instantaneous* polarization and rotation angle of a beam. Causing that vector to trace a trajectory over time on the surface of the sphere represents a *polarization-modulated* (and *rotation-modulated*) beam [Fig. 3.1(B)]. If, then, the beam is sampled by a device at a rate which is less than the rate of modulation, the sampled output from the device will be a condensation of *two components* of the wave, which are continuously changing with respect to each other, into *one snapshot* of the wave, at *one location* on the surface of the sphere and *one instantaneous polarization and axis rotation*. Thus, from the viewpoint of a device sampling at a rate less than the modulation rate, a two-to-one mapping (over time) has occurred, which is the signature of an SU(2) field.

The modulations which result in trajectories on the sphere are infinite in number. Moreover, those modulations, at a rate of multiples of 2π greater than 1, which result in the return to a single location on the sphere at a frequency of exactly 2π, will all be detected by the device sampling at a rate of 2π as the same. In other words, the device cannot detect what kind of simple or complicated trajectory was performed between departure from, and arrival at, the same location on the sphere. To the relatively slowly sampling device, the fast modulated beam can have "internal energies" quite unsuspected.

We can say that such a static device is *a unipolar, set-rotational-axis, U(1) sampling device* and the fast polarization (and rotation)-modulated beam is a *multipolar, multirotation axis, SU(2) beam*.

[ddd]Based on: Barrett, T.W., "On the distinction between fields and their metric: the fundamental difference between specifications concerning medium-independent fields and constitutive specifications concerning relations to the medium in which they exist," *Annales de la Fondation Louis de Broglie*, vol. 14, pp. 37–75, 1989.
[eee]Poincaré, H., *Théorie Mathématique de la Lumière* (Georges Carré, Paris, 1892), vol. 2, chap. 2.

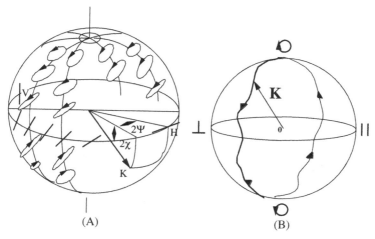

Fig. 3.1. (A) Poincaré sphere representation of wave polarization and rotation. (B) A Poincaré sphere representation of signal polarization (longitudinal axis) and polarization-rotation (latitudinal axis). A representational trajectory of polarization-rotation modulation is shown by changes in the vector at the center of the sphere and pointing at the surface. Waves of various polarization modulations, $\partial\phi^n/\partial t^n$, can be represented as trajectories on the sphere. The case shown is an arbitrary trajectory repeating 2π. (After Ref. ggg.)

The reader may ask: How many situations are there in which a sampling device, at set unvarying polarization, samples at a slower rate than the modulation rate of a radiated beam? The answer is that there is an infinite number, because nature is set up to be that way.[fff]   For example, the period of modulation can be faster than the electronic or vibrational or dipole relaxation times of any atom or molecule. In other words, *pulses or wave packets* (which, in temporal length, constitute the sampling of a continuous wave, continuously polarized and rotation-modulated, but sampled only over a temporal length between arrival and departure time at the instantaneous polarization of the sampler of set polarization and rotation — in this case an electronic

---

[fff]See Barrett, T.W., "Is quantum physics a branch of sampling theory?" *Courants, amers, écueils en microphysique*, C. Cormier-Delanoue, G. Lochak, P. Lochak (eds.), (Fondation Louis de Broglie, 1993).

[ggg]Barrett, T.W., "Polarization-rotation modulated, spread polarization-rotation, wide-bandwidth radio-wave communications system," United States Patent 5,592,177, dated Jan. 7, 1997.

or vibrational state or dipole) have an internal modulation at a rate greater than that of the relaxation or absorption time of the electronic or vibrational state.

The representation of the sampling by a unipolar, single-rotation-axis, U(1) sampler of an SU(2) continuous wave which is polarization- and rotation-modulated is shown in Fig 3.2, which is the correspondence between the output space sphere and an Argand plane.[hhh] The Argand plane, Σ, is drawn in two dimensions, x and y, with $z = 0$, and for a set snapshot in time. A point on the Poincaré sphere is represented as $P(t, x, y, z)$, and as in this representation $t = 1$ (or one step in the future), specifically as $P(1, x, y, z)$. The Poincaré sphere is also identified as a three-sphere, S^+, which is defined in Euclidean space as

$$x^2 + y^2 + z^2 = 1.$$

The sampling described above is represented as a mapping of a point $P(1, x, y, z)$ in S^+, and of SU(2) symmetry, to a point $P'(1, x', y', z')$ on Σ, and in U(1) symmetry.

The point P' can then be labeled by a single complex parameter:

$$\zeta = X' + iY'.$$

Using the definition

$$z = 1 - \frac{CA}{CP'} = 1 - \frac{NP}{NP'} = 1 - \frac{NB}{NC},$$

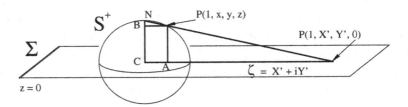

Fig. 3.2. Correspondence between the output space sphere and an Argand plane. (After Penrose and Rindler.[30])

[hhh] After Penrose, R. & Rindler, W., *Spinors and Space–Time, Volume 1, Two-Spinor, Calculus and Relativistic Fields* (Cambridge University Press, 1984).

then

$$\zeta = \frac{x + iy}{1 - z}.$$

A pair (ξ, η) of complex numbers can be defined as

$$\zeta = \frac{\xi}{\eta},$$

and Penrose and Rindler[28] have shown, in another context, that what we have identified as the *presampled* SU(2) polarization- and rotation-modulated wave can be represented in the units of

$$W = \frac{1}{\sqrt{2}} (\xi\xi^* + \eta\eta^*),$$

$$X = \frac{1}{\sqrt{2}} (\xi\eta^* + \eta\xi^*),$$

$$Y = \frac{1}{i\sqrt{2}} (\xi\eta^* - \eta\xi^*),$$

$$Z = \frac{1}{\sqrt{2}} (\xi\xi^* - \eta\iota^*).$$

These definitions make explicit that a complex linear transformation of the U(1) ξ and η results in a real linear transformation of the SU(2) (W, X, Y, Z).

Therefore, a complex linear transformation of ξ and η can be defined:

$$\xi \mapsto \xi' = \alpha\xi + \beta\eta, \tag{3.1a}$$

$$\eta \mapsto \eta' = \gamma\xi + \delta\eta,$$

$$\text{or} \quad \zeta \mapsto \zeta' = \frac{\alpha\zeta + \beta}{\gamma\zeta + \delta} \tag{3.1b}$$

where α, β, γ and δ are arbitrary nonsingular complex numbers.

Now the transformations, 3.1(a) and 3.1(b), are *spin transformations*, implying that

$$\zeta = \frac{X + iY}{T - Z} = \frac{W + Z}{X - iY},$$

and if a spin matrix, A, is defined as

$$A = \begin{pmatrix} \alpha & \beta \\ \gamma & \delta \end{pmatrix}, \quad \det A = 1,$$

then the two transformations, 3.1(a), are

$$A = \begin{pmatrix} \xi' & \xi \\ \eta' & \eta \end{pmatrix},$$

which means that the spin matrix of a composition is given by the product of the spin matrix of the factors. Any transformation of the (3.2) form is linear and real and leaves the form $W^2 - X^2 - Y^2 - Z^2$ invariant.

Furthermore, there is a unimodular condition,

$$\alpha\delta - \beta\gamma = 1,$$

and the matrix A has the inverse

$$A^{-1} = \begin{pmatrix} \delta & -\beta \\ -\gamma & \alpha \end{pmatrix},$$

which means that the spin matrix A and its inverse A^{-1} gives rise to the same transformation of ζ even though they define different spin transformations. Owing to the unimodular condition, the A spin matrix is unitary or

$$A^{-1} = A^*,$$

where A^* is the conjugate transpose of A.

The consequence of these relations is that *every proper 2π rotation on S^+ — in the present instance the Poincaré sphere — corresponds to precisely two unitary spin rotations.* As every rotation on the Poincaré sphere corresponds to a polarization-rotation *modulation, every proper 2π polarization-rotation modulation corresponds to precisely two unitary spin rotations.* The vector K in Fig. 3.1 corresponds to two vectorial components, one being the negative of the other. As every unitary spin transformation corresponds to a unique proper rotation of S^+, any *static* (unipolarized, e.g. linearly, circularly or elliptically polarized, as opposed to polarization-modulated) representation on S^+ (Poincaré sphere) corresponds to a trisphere representation [Fig. 3.3(A)]. Therefore

$$A^{-1}A = \pm I,$$

where I is the identity matrix. *Thus, a spin transformation is defined uniquely up to sign by its effect on a static instantaneous snapshot*

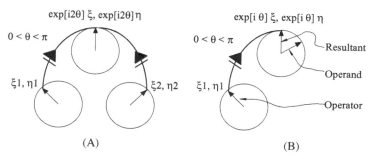

exp[i2θ] ξ, exp[i2θ]η

$0 < \theta < \pi$

$\xi1, \eta1$

$\xi2, \eta2$

exp[i θ] ξ, exp[i θ] η

$0 < \theta < \pi$

Resultant

Operand

$\xi1, \eta1$

Operator

(A)

(B)

Fig. 3.3. (A) Trisphere representation of *static polarization mapping*: $\xi_1, \eta_1; \xi_2 \eta_2 \mapsto e^{i2\theta}\xi; e^{i2\theta}\eta; 0 < \theta < \pi$. Note that a 360° excursion of $\xi_1\eta_1$ and $\xi_2\eta_2$ corresponds to a 360° excursion of $e^{i2\theta}\xi, e^{i2\theta}\eta$, i.e. this is a mapping for static polarization. (B) Bisphere representation of *polarization modulation mapping* (or $\xi_1\eta_1 \mapsto e^{i\theta}\xi, e^{i\theta}\eta; 0 < \theta < \pi$) exhibiting the property of spinors that, corresponding to two unitary transformations of e.g. 2π, i.e. 4π, a null rotation of 2π is obtained. Notice that for a 360° rotation of the resultant (i.e. the final output wave), and with a stationary operand, the operator must be rotated through 720°.

representation on the S^+ (Poincaré) sphere:

$$\xi_1, \eta_1; \xi_2, \eta_2 \mapsto e^{i2\theta}\xi, e^{i2\theta}\eta; \quad 0 < \theta < \pi.$$

Turning now to the case of polarization-rotation modulation, or *continuous rotation* of $\xi_1\eta_1; \xi_2\eta_2$: corresponding to a continuous rotation of $\xi_1\eta_1; \xi_2\eta_2$ through 2θ, there is a rotation of the resultant through θ. This correspondence is a consequence of the $A^{-1}A = \pm I$ relation; namely, that if the unitary transformation of A or A^{-1} is applied separately, the identity matrix will *not* be obtained. However, if the unitary transformation is applied *twice*, then the identity matrix *is* obtained; and from this follows the remarkable properties of spinors that corresponding to two unitary transformations of e.g. 2π, i.e. 4π, one null vector rotation of 2π is obtained. This is a bisphere correspondence and is shown in Fig. 3.3(B). This figure also represents the case of polarization-rotation modulation — as opposed to static polarization-rotation.

We now identify the vector, K, in Fig. 3.1 as a *null vector* defined as

$$K = Ww + Xx + Yy + Zz,$$

the coordinates of which satisfy

$$W^2 - X^2 - Y^2 - Z^2 = 0,$$

where W, X, Y and Z are functions of time: $W(t)$, $X(t)$, $Y(t)$ and $Z(t)$. The distinguishing feature of this null vector is that phase transformations $\xi \mapsto e^{i\theta}\xi$, $\eta \mapsto e^{i\theta}\eta$ leave K unchanged, i.e. K represents ξ and η only up to phase — which is the hallmark of a U(1) representation.

K thus defines a static polarization-rotation — whether linear, circular or elliptical — on the Poincaré sphere. The ξ, η representation of the vector K *gives no indication of the future position of K*, i.e. the representation does not address the indicated hatched trajectory of the vector K around the Poincaré sphere. But it is precisely this trajectory which defines the particular polarization modulation for a specific wave. *Stated differently: a particular position of the vector K on the Poincaré sphere gives no indication of its next position at a later time*, because the vector can depart (be joined) in any direction from that position when only the static ξ, η coordinates are given.

In order to address *polarization-rotation modulation* — not just static polarization-rotation — an algebra is required which can reduce the ambiguity of a static representation. Such an algebra which is associated with ξ, η, and which reduces the ambiguity up to a sign ambiguity, is available in the twistor formalism.[iii] In this formalism, polarization-rotation modulation can be accommodated, and a spinor, κ, can be represented not only by a null direction indicated by ξ, η or ζ, but also a real tangent vector L, indicated in Fig. 3.4.

Using this algebraic formalism, the Poincaré vector — *and its direction of change (up to a sign ambiguity)* — can be represented. A real tangent vector L of S^+ at P is defined as

$$L = \frac{\lambda \partial}{\partial \zeta} + \frac{\lambda^* \partial}{\partial \zeta^*},$$

[iii]Penrose, R. & Rindler, W., *Spinors and Space–Time, Volume 1, Two-Spinor Calculus and Relativistic Fields* (Cambridge University Press, 1984).

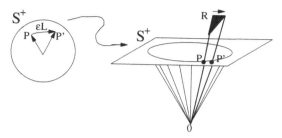

Fig. 3.4. Relation of a trajectory in a specific direction on an output sphere S^+ and a null flag representation on the hyperplane, W, intersection with S^+. (After Penrose and Rindler.[iii])

where λ is some expression in ξ, η. With the choice $\lambda = -\left(\frac{1}{\sqrt{2}}\right)\eta^{-2}$

$$L = \left(\frac{1}{\sqrt{2}}\right)\left[\eta^{-2}\left(\frac{\partial}{\partial\zeta}\right) + \eta^{*-2}\left(\frac{\partial}{\partial\zeta^*}\right)\right],$$

and thus knowing L at P (as an operator) means that the pair ξ, η is known completely up to sign, or, for any $f(\zeta, \zeta^*)$,

$$\frac{1}{\varepsilon_{\lim \varepsilon \to 0}}\left(fp' - fp\right) = Lf.$$

Succinctly: the tangent vector L in the abstract space S^+ (Poincaré sphere) corresponds to a tangent vector L in the coordinate-dependent representation S^+ of S^+. L is a unit vector if and only if K, the null vector corresponding to ξ, η, defines a point actually on S^+. Therefore, a plane of K and L can be defined by

$$aK + bL,$$

and if $b > 0$, then a half-plane, Π, is defined bounded by K. K and L are both spacelike and orthogonal to each other. In the twistor formalism, Π and K are referred to as a *null flag* or a *flag*. The vector K is called the *flagpole*, its direction is the *flagpole direction* and the half-plane, Π, is the *flag plane*.

Our conclusions are that a polarization-rotation-modulated wave can be represented as a periodic trajectory of polarization-rotation modulation on a Poincaré sphere, or a spinorial object. A defining characteristic of a spinorial object is that it is not returned to its original state when rotated through an angle 2π about some axis, but

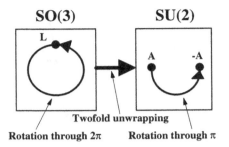

Fig. 3.5. The left side [SO(3)] describes the symmetry of the trajectory K on the Poincaré sphere; the right side describes the symmetry of the associated $Q_1(\psi, \chi)$ and $Q_2(\psi, \chi)$ which are functions of the ψ, χ angles on the Poincaré sphere. (Adapted from Penrose and Rindler.)

only when rotated through 4π. Referring to Fig. 3.5, we see that for the resultant to be rotated through 2π and returned to its original polarization state, the *operator* must be rotated through 4π. Thus a spinorial object (polarization-rotation-modulated beams) exists in a different topological space from static polarized-rotated beams due to the additional degree of freedom provided by the polarization bandwidth, which does not exist prior to modulation.

For example, let us consider constituent polarization vectors, $Q^i(\omega, \delta)$, and let C be the space orientations of $Q^i(\omega, \delta)$. A spinorized version of $Q^i(\omega, \delta)$ can be constructed provided the space is such that it possesses a twofold universal covering space C^*, and provided the two different images, $Q_1(\psi, \chi)$ and $Q_2(\psi, \chi)$ existing in C^* of an element existing in C, are interchanged after a continuous rotation through 2π is applied to a $Q^i(\omega, \delta)$. In the case we are considering, C has the topology of the SO(3) group, but C^* of the SU(2) group (which is the same as the space of unit quaternions) Thus, there is a $2 \to 1$ relation between the SO(3) object and the SU(2) object (Fig. 3.5).

We may take $Q^i(\omega, \delta)$ to be polarization vectors (null flags) and C to be the space of null flags. The spinorized null flags, $Q_1(\psi, \chi)$ and $Q_2(\psi, \chi)$, are elements of C^*, i.e. they are spin vectors. Referring to Figs. 3.3(B) and 3.5, we see that each null flag, $Q^i(\omega, \delta)$, defines two associated spin vectors, κ and $-\kappa$. A continuous rotation through 2π will carry κ into $-\kappa$ by acting on (ξ, η). On repeating the process, $-\kappa$

is carried back into κ:

$$-(-\kappa) = \kappa.$$

Furthermore, any spin vector, κ_1, can be represented as a linear combination of two spin vectors κ_2 and κ_3:

$$\{\kappa_2, \kappa_3\}\kappa_1 + \{\kappa_3, \kappa_1\}\kappa_2 + \{\kappa_1, \kappa_2\}\kappa_3 = 0,$$

where { } indicates the antisymmetrical inner product. Thus any arbitrary polarization can be represented as a linear combination of spin vectors.

A generalized representation of spin vectors (and thus of polarization-rotation modulation) that is in terms of components is obtained using a normalized pair, a,b, as a spin frame:

$$\{a, b\} = -\{b, a\} = 1.$$

Therefore

$$\kappa = \kappa^0 a + \kappa^1 b,$$

with

$$\kappa^0 = \{\kappa, b\}, \quad \kappa^1 = -\{\kappa, a\}.$$

The flagpole of a is $\dfrac{t+z}{\sqrt{2}}$ and of b is $\dfrac{t-z}{\sqrt{2}}$ and can be represented over time in Minkowski tetrad (t, x, y, z) form (t_1 representation) and for multiple time frames or sampling intervals providing overall $(t_1 \cdots t_n)$ a Cartan–Weyl form representation (Fig. 3.6) by using sampling intervals which "reset the clock" after every sampling of instantaneous polarization. Thus polarization modulation is represented by the continuous changes in a, b over time or the collection of samplings of a, b over time as depicted in Fig. 3.6.

The relation to the electromagnetic field is as follows. The (antisymmetrical) inner product of two spin vectors can represented as

$$\{\kappa_1, \kappa_2\} = \varepsilon_{AB}\kappa^A\kappa^B = -\{\kappa_2, \kappa_1\},$$

where the ε (or the fundamental numerical metric spinors of second rank) are antisymmetrical:

$$\varepsilon_{AB}\varepsilon^{CB} = -\varepsilon_{AB}\varepsilon^{BC} = \varepsilon_{AB}\varepsilon^{BC} = -\varepsilon_{BA}\varepsilon^{CB} = \varepsilon_A^C = -\varepsilon_A^C,$$

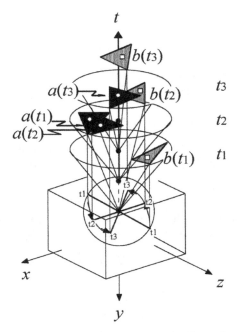

Fig. 3.6.   Spin frame representation of a spin vector by flag pole normalized pair representation $\{a, b\}$ over the Poincaré sphere in Minkowski tetrad (t, x, y, z) form (t_1 representation) and for three time frames or sampling intervals providing overall ($t_1 \cdots t_n$) a Cartan–Weyl form representation. The sampling intervals "reset the clock" after every sampling of instantaneous polarization. Thus polarization modulation is represented by the collection of samplings over time. [Minkowski form after Penrose and Rindler (1984)]. This is an SU(2) $Q_i(\psi, \chi)$ in C^* representation, *not* an SO(3) $Q^i(\omega, \delta)$ in C representation over 2π. This can be seen by noting that $a \longmapsto b$ or $b \longmapsto a$ over π, not 2π, while the polarization modulation in SO(3) repeats at a period of 2π.

with a canonical mapping (or isomorphism) between, for example, κ^B and κ_B:

$$\kappa^B \mapsto \kappa_B = \kappa^A \varepsilon_{AB}.$$

A potential can be defined:

$$\Phi_A = i(\varepsilon\alpha)^{-1}\nabla_A\alpha,$$

where α is a gauge

$$\alpha\alpha^* = 1,$$

and ∇_A is a covariant derivative, $\partial/\partial x^A$, but without the commutation property. The covariant electromagnetic field is then

$$F_{AB} = \nabla_A \Phi_B - \nabla_B \Phi_A + ig \left[\Phi_B, \Phi_A \right],$$

where g is generalized charge.

A physical representation of the polarization modulated [SU(2)] beam can be obtained using a Lissajous pattern[iii] representation (Figs. 3.7–3.9).

The controlling variables for polarization and rotation modulation are given in Table 3.1. We can note that the Stokes parameters (s_0, s_1, s_2, s_3) defined over the SU(2) dimensional variables, ψ, χ, of $Q_i(\psi, \chi)$ are sufficient to describe *polarization-rotation-modulation*, and relate those variables to the SO(3) dimensional variables, $\omega(\tau, z)$, δ, of $Q^i(\omega, \delta)$, which are sufficient to describe the *static*

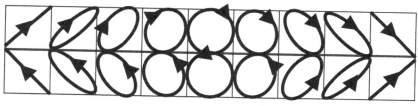

Fig. 3.7. Lissajous patterns representing a polarization-modulated electric field over time, viewed in the plane of incidence, resulting from the two orthogonal s and p fields, which are out of phase by the following degrees: 0, 21, 42, 64, 85, 106, 127, 148, 169 (top row); 191, 212, 233, 254, 275, 296, 318, 339, 360 (bottom row). In these Lissajous patterns, the plane polarizations are represented at 45° to the axes. In this example, there is a simple constant rate polarization with no rotation modulation. This is an SO(3) $Q^i(\omega, \delta)$ in C representation over 2π, *not* an SU(2) $Q_i(\psi, \chi)$ in C* over π.

[iii]Lissajous patterns are the locus of the resultant displacement of a point which is a function of two (or more) simple periodic motions. In the usual situation, the two periodic motions are orthogonal (i.e. at right angles) and are of the same frequency. The Lissajous figures then represent the polarization of the resultant wave as a diagonal line: top left to bottom right in the case of linear perpendicular polarization; bottom left to top right in the case of linear horizontal polarization; a series of ellipses, or a circle, in the case of circular corotating or contrarotating polarization; all of these corresponding to the possible differences in constant phase between the two simple periodic motions. If the phase is not constant, but is changing or modulated, as in the case of polarization modulation, then the pattern representing the phase is constantly changing over the time the Lissajous figure is generated. [Named after Jules Lissajous (1822–1880).]

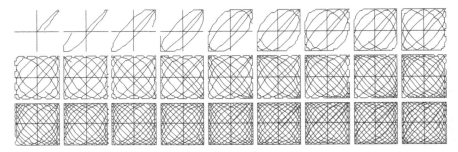

Fig. 3.8. Lissajous patterns representing the polarized electric field over time, viewed in the plane of incidence resulting from the two orthogonal s and p fields. The p field is phase-modulated at a rate $d\phi/dt = 0.2\,t$. In these Lissajous patterns, the plane polarizations are represented at 45° to the axes. This is an SO(3) $Q^i(\omega, \delta)$ in C representation over 2π, *not* an SU(2) $Q_i(\psi, \chi)$ in C* over π.

polarization-rotation conditions of linear, circular, left- and right-handed polarization-rotation.

We can also note the fundamental role that concepts of topology played in distinguishing *static polarization-rotation* from *polarization-rotation modulation*.

4. The Aharonov–Bohm Effect

We consider again the Aharonov–Bohm effect as an example of a phenomenon understandable only from topological considerations. To summarize the description in Chapter 1: Beginning in 1959, Aharonov and Bohm[kkk] challenged the view that the classical vector potential produces no observable physical effects by proposing two experiments. The one which is most discussed is shown in Fig. 4.1, which is Fig. 3.1.1 of Chapter 1, provided here for convenience. A beam of monoenergetic electrons exists from a source at X and is diffracted into two beams by the slits in a wall at Y1 and Y2. The two beams produce an interference pattern at III which is measured. Behind the wall is a solenoid, the **B** field of which points out of the paper. The

---

[kkk] Aharonov, Y. & Bohm, D., "Significance of the electromagnetic potentials in quantum theory," *Phys. Rev.*, vol. 115, pp. 485–491, 1959.

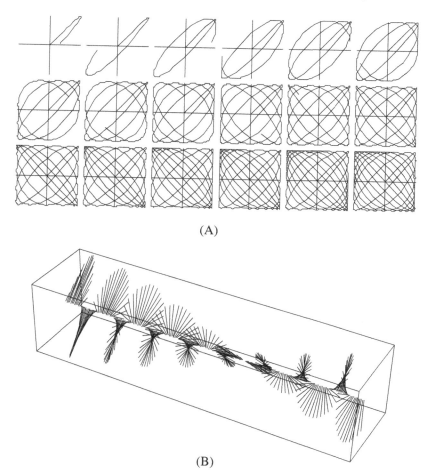

(A)

(B)

Fig. 3.9. (A) Lissajous patterns representing the polarized electric field over time, viewed in the plane of incidence, resulting from the two orthogonal s and p fields, which are out of phase by the following degrees: 0, 21, 42, 64, 85, 106, 127, 148, 169 (top row); 191, 212, 233, 254, 275, 296, 319, 339, 360 (bottom row). In these Lissajous patterns, the plane polarizations are represented at 45° to the axes. (B) Representation of a polarization-modulated beam over $2p$ in the z direction. These are SO(3) $Q^i(\omega, \delta)$ in C representations over 2π, *not* an SU(2) $Q_i(\psi, \chi)$ in C^* over π.

Table 3.1. Controlling variables.[III]

Field input variables (coordinate axes)	$E_x = a_1 \cos(\tau + \delta_1)$ $E_y = a_2 \cos(\tau + \delta_2)$ $\tau = \omega t - \kappa z$
Field input variables (ellipse axes)	$E_\xi = a \cos(\tau + \delta) = E_x \cos \psi + E_y \sin \psi$ $E_\eta = \pm b \cos(\tau + \delta) = -E_x \sin \psi + E_y \cos \psi$ $\tau = \omega t - \kappa z$
Phase variables	$\delta = \delta_2 - \delta_1;$ $\left(\frac{E_x}{a_1}\right)^2 + \left(\frac{E_y}{a_2}\right)^2 - 2\frac{\cos \delta}{a_1 a_2} = \sin^2 \delta$
Auxiliary angle, α	$\frac{a_2}{a_1} = \tan(\alpha)$
Control variables	$a_1, a_2, \delta_1, \delta_2$
Resultant transmitted variables and relation of coordinate axes, a_1, a_2, to ellipse axes, a, b	$a^2 + b^2 = a_1^2 + a_2^2$
Rotation	$\tan(2\psi) = (\tan(2\alpha)) \cos(\delta) = \frac{2a_1 a_2}{a_1^2 - a_2^2} \cos \delta$
Ellipticity	$\sin(2\chi) = (\sin(2\alpha)) \sin(\delta); \tan(\chi) = \pm \frac{b}{a}$
Rotation	ψ — resultant determined by a_1 and a_2 with δ constant
Ellipticity	χ — resultant determined by δ with a_1 and a_2 constant
Determinant of rotation ψ	a_1, a_2 with δ constant
Determinant of ellipticity χ	δ with a_1, a_2 constant
Stokes parameters	$s_0 = a_1^2 + a_2^2$ $s_1 = a_1^2 - a_2^2 = s_0 \cos(2\chi) \cos(2\psi)$ $s_2 = 2a_1 a_2 \cos(\delta) = s_0 \cos(2\chi) \sin(2\psi) = s_1 \tan(2\psi)$ $s_3 = 2a_1 a_2 \sin(\delta) = s_0 \sin(2\chi)$
Linear polarization condition	$\delta = \delta_2 - \delta_1 = m\pi,$ $m = 0, \pm 1, \pm 2, \ldots$ $\frac{E_y}{E_x} = (-1)^m \frac{a_2}{a_1}$
Circular polarization condition	$a_1 = a_2 = a;$ $\delta = \delta_2 - \delta_1 = \frac{m\pi}{2}$, $m = \pm 1, \pm 3, \pm 5, \ldots$ $E_x^2 + E_y^2 = a^2$
Right-hand polarization condition	$\sin \delta > 0,$ $\delta = \frac{\pi}{2} + 2m\pi,$ $m = 0, \pm 1, \pm 2, \ldots$ $E_x = a \cos(\tau + \delta_1)$ $E_y = a \cos(\tau + \delta_1 + \frac{\pi}{2}) = -a \sin(\tau + \delta_1)$
Left-hand polarization condition	$\sin \delta < 0,$ $\delta = -\frac{\pi}{2} + 2m\pi,$ $m = 0, \pm 1, \pm 2, \ldots$ $E_x = a \cos(\tau + \delta_1)$ $E_y = a \cos(\tau + \delta_1 - \frac{\pi}{2}) = a \sin(\tau + \delta_1)$

[III]After Born, M. & Wolf, E., *Principles of Optics*, 7th edn. (Cambridge University Press, 1999).

postulate of an absence of a free local magnetic monopole in conventional U(1) electromagnetism ($\nabla \cdot \mathbf{B} = 0$) predicts that the magnetic field outside the solenoid is zero. Before the current is turned on in the solenoid, there should be the usual interference patterns observed at III, of course, due to the differences in the two path lengths.

Aharonov and Bohm made the important prediction that if the current is turned on, then due to the differently directed $\mathbf{A}$ fields along paths 1 and 2 indicated by the arrows in Fig. 4.1, additional phase shifts should be discernible at III. This prediction was confirmed experimentally[mmm] and the evidence for the effect has been extensively reviewed.[nnn]

It is the opinion of this book that the topology of this situation is fundamental and dictates its explanation. Therefore, we must clearly note the topology of the physical layout of the design of the situation which exhibits the effect. The physical situation is that of an *interferometer*.

[mmm]Chambers, R.G., "Shift of an electron interference pattern by enclosed magnetic flux," *Phys. Rev. Lett.*, vol. 5, pp. 3–5, 1960;

Boersch, H., Hamisch, H., Wohlleben, D. & Grohmann, K., "Antiparallele Weissche Bereiche als Biprisma für Elektroneninterferenzen," *Zeitschrift für Physik*, vol. pp. 159, 397–404, 1960;

Möllenstedt, G. & Bayh, W., Messung der kontinuierlichen Phasenschiebung von Elektronenwellen im kraftfeldfreien Raum durch das magnetische Vektorpotential einer Luftspule," *Die Naturwissenschaften*, vol. 49, pp. 81–82, 1962;

Matteucci, G. & Pozzi, G., "New diffraction experiment on the electrostatic Aharonov–Bohm effect," *Phys. Rev. Lett.*, vol. 54, pp. 2469–2472, 1985;

Tonomura, A., *et al.*, "Observation of Aharonov–Bohm effect by electron microscopy," *Phys. Rev. Lett.*, vol. 48, pp. 1443–1446, 1982;

———, "Is magnetic flux quantized in a toroidal ferromagnet?" *Phys. Rev. Lett.*, vol. 51, pp. 331–334, 1983;

———, "Evidence for Aharonov–Bohm effect with magnetic field completely shielded from electron wave," *Phys. Rev. Lett.*, vol. 56, pp. 792–795, 1986;

——— & Callen, E., "Phase, electron holography and conclusive demonstration of the Aharonov–Bohm effect. *ONRFE Sci. Bul.*, vol. 12, no. 3, pp. 93–108, 1987.

[nnn]Berry, M.V., "Exact Aharonov–Bohm wavefunction obtained by applying Dirac's magnetic phase factor" *Eur. J. Phys.*, vol. 1, pp. 240–244, 1980;

Peshkin, M., "The Aharonov–Bohm effect: why it cannot be eliminated from quantum mechanics" *Phys. Rep.*, vol. 80, pp. 375–386, 1981;

Olariu, S. & Popescu, I.I., "The quantum effects of electromagnetic fluxes," *Rev. Mod. Phys.*, vol. 157, pp. 349–436, 1985;

Horvathy, P.A., "The Wu–Yang factor and the non-Abelian Aharonov–Bohm experiment," *Phys. Rev.*, vol. D33, pp. 407–414, 1986;

Peshkin, M. & Tonomura, A., *The Aharonov–Bohm Effect* (Springer-Verlag, New York, 1989).

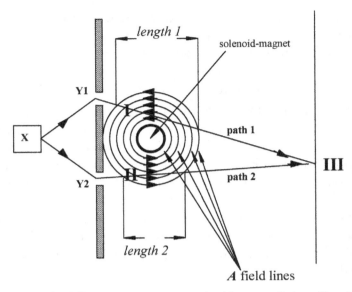

Fig. 4.1. Two-slit diffraction experiment on the Aharonov–Bohm effect. Electrons are produced by a source at X, pass through the slits of a mask at Y1 and Y2, interact with the *A* field at locations I and II over lengths l_1 and l_2, respectively, and their diffraction pattern is detected at III. The solenoid magnet is between the slits and is directed out of the page. The different orientations of the external *A* field at the places of interaction I and II of the two paths 1 and 2 are indicated by arrows following the right-hand rule.

That is, there are two paths around a central location — occupied by the solenoid — and a measurement is taken at a location, III, in the Fig. 4.1, where there is overlap of the wave functions of the test waves which have traversed, separately, the two different paths. (The test waves or test particles are complex wave functions with phase.) It is important to note that the overlap area, at III, is the only place where a measurement can take place of the effects of the **A** field [which occurred earlier and at other locations (I and II)]. The effects of the **A** field occur along the two different paths and at locations I and II, but they are *inferred*, and not measurable there. Of crucial importance in this special interferometer is the fact that the solenoid presents a *topological obstruction*. That is, if one were to consider the two joined paths of the interferometer as a raceway or a loop and one squeezed the loop tighter and tighter, then nevertheless one could not in this situation—unlike in most situations — reduce the interferometer's raceway of paths to a single point. (Another way of saying this is: not all closed curves in a region need to have a vanishing line integral, because one exception is a loop with an obstruction.) The reason one cannot reduce the interferometer to a single point is that there exists in its middle the solenoid, which is a positive quantity, and acts as an obstruction.

It is the view of this book that the existence of the obstruction changes the situation entirely. *Without* the existence of the solenoid in the interferometer, the loop of the two paths *can be* reduced to a single point and the region occupied by the interferometer is then *simply connected*. But *with* the existence of the solenoid, the loop of the two paths *cannot be* reduced to a single point and the region occupied by this special interferometer is then *multiply connected*. The Aharonov–Bohm effect exists only in the scenario of multiply-connectedness. But we should note that the Aharonov–Bohm effect is a *physical* effect and simply- and multiply-connectedness are *mathematical descriptions* of physical situations.

The topology of the physical interferometric situation addressed by Aharonov and Bohm defines the physics of that situation and also the mathematical description of that physics. If that situation *were not* multiply connected, but simply connected, then there would be no interesting physical effects to describe. The situation would be described by U(1) electromagnetics and the mapping from one region to another is conventionally one-to-one. However, as the Aharonov–Bohm situation is multiply connected, there is a two-to-one mapping [SU(2)/Z_2] of the two different regions of the two paths to the single region at III where a measurement is made. Essentially, at III a measurement is made of *the differential histories* of the *two* test waves which traversed the *two* different paths and experienced *two* different forces resulting in two different phase effects.

In conventional, i.e. normal U(1) or simply connected situations, the fact that a vector field, viewed axially, is pointing in one direction, if penetrated from one direction on one side, and is pointing in *the opposite direction*, if penetrated from the same direction, but *on the other side*, is of no consequence at all — because that field is of U(1) symmetry and can be reduced to a single point. Therefore, in most cases which are of U(1) symmetry, we do not need to distinguish the direction of the vectors of a field from one region to another of that field. However, the Aharonov–Bohm situation is not conventional or simply connected, but special. (In other words, the physical situation associated with the Aharonov–Bohm effect has a nontrivial topology.) It is a multiply connected situation and of SU(2)/Z_2 symmetry. Therefore the direction of the *A* field on the separate paths is of crucial importance, because a test wave traveling along one path will experience an **A** vectorial component directed *against* its trajectory and thus be retarded, and another test wave traveling along another path will experience an **A** vectorial component directed *with* its trajectory and thus its speed is boosted. These "retardations" and "boostings" can be measured as phase changes, *but not at the time or at the locations I and II, where their occurrence is separated along the two different paths,* but *later,* and at the *overlap location* III. It is important to note that if measurements are attempted at locations I and II in Fig. 4.1, these effects will not be seen because there is no two-to-one mapping at either I on II and therefore no referents. The locations I or II are both simply connected with the source and therefore only the conventional U(1) electromagnetics applies at these locations (with respect to the source). It is only region III which is multiply connected with the source and at which the histories of what happened to the test particles at I and II can be measured. In order to distinguish the "boosted" **A** field (because the test wave is traveling "with" its direction) from the "retarded" **A** field (because the test wave is traveling "against" its direction), we introduce the following notation: $\mathbf{A}_+$ and $\mathbf{A}_-$.

Because of the distinction between the **A** oriented potential fields at positions I and II — which *are not* measurable and are *vectors or numbers* of U(1) symmetry — and the *A* potential fields at III — which *are* measurable and are *tensors or matrix-valued functions* of (in the present instance) SU(2)/Z_2 = SO(3) symmetry (or higher symmetry) — for reasons of clarity we might introduce a distinguishing notation. In the case of the potentials of U(1) symmetry at I and II, we might use the lower case a_μ, $\mu = 0, 1, 2, 3$, and for the potentials of SU(2)/Z_2 = SO(3) at III we might use the upper case $\mathbf{A}_\mu$,

$\mu = 0, 1, 2, 3$. Similarly, for the electromagnetic field tensor at I and II, we might use the lower case $f_{\mu\nu}$, and for the electromagnetic field tensor at III, we might use the upper case $F_{\mu\nu}$. Then the definitions for the electromagnetic field tensor are:

At locations I and II the Abelian relationship is

$$f_{\mu\nu}(x) = \partial_\nu a_\mu(x) - \partial_\mu a_\nu(x), \tag{4.1}$$

where, as is well known, $f_{\mu\nu}$ is Abelian and gauge-*invariant*;

But at location III the non-Abelian relationship is

$$F_{\mu\nu} = \partial_\nu \mathbf{A}_\mu(x) - \partial_\mu \mathbf{A}_\nu(x) - ig_m \left[\mathbf{A}_\mu(x), \mathbf{A}_\nu(x) \right], \tag{4.2}$$

where $F_{\mu\nu}$ is gauge-*covariant*, g_m is the magnetic charge density and the brackets are commutation brackets. We remark that in the case of non-Abelian groups, such as SU(2), the potential field *can carry charge*. It is important to note that if the physical situation changes from SU(2) symmetry back to U(1), then $F_{\mu\nu} \rightarrow f_{\mu\nu}$.

Despite the clarification offered by this notation, the notation can also cause confusion, because in the present literature, the electromagnetic field tensor is *always* referred to as F, whether F is defined with respect to U(1) or SU(2) or other symmetry situations. Therefore, although we prefer this notation, we shall not proceed with it. However, it is important to note that the A field in the U(1) situation is a *vector or a number*, but in the SU(2) or non-Abelian situation, it is a *tensor or a matrix-valued function*.

We referred to the physical situation of the Aharonov–Bohm effect as an interferometer around an obstruction and it is two-dimensional. It is important to note that the situation is not provided by a toroid, although a toroid is also a physical situation with an obstruction and the fields existing on a toroid are also of SU(2) symmetry. However, the toroid provides a two-to-one mapping of fields in not only the x and y dimensions but also the z dimension, and *without* the need of an electromagnetic field pointing in two directions, $+$ and $-$. The physical situation of the Aharonov–Bohm effect is defined only in the x and y dimensions (there is no z dimension) and, in order to be of SU(2)/Z_2 symmetry, *requires* a field to be oriented differentially on the separate paths. If the differential field is removed from the Aharonov–Bohm situation, then that situation reverts to a simple interferometric raceway which can be reduced to a single point and with no interesting physics.

How does the topology of the situation affect the explanation of an effect? A typical previous explanation[ooo] of the Aharonov–Bohm effect commences with the Lorentz force law:

$$\mathscr{F} = e\mathbf{E} + ev \times \mathbf{B}. \tag{4.3}$$

The electric field, $\mathbf{E}$, and the magnetic flux density, $\mathbf{B}$, are essentially confined to the inside of the solenoid and therefore cannot interact with the test electrons. An argument is developed by defining the $\mathbf{E}$ and $\mathbf{B}$ fields in terms of the $\mathbf{A}$ and ϕ potentials:

$$\mathbf{E} = -\frac{\partial \mathbf{A}}{\partial t} - \nabla\phi, \quad \mathbf{B} = \nabla \times \mathbf{A}. \tag{4.4}$$

Now we can note that these conventional U(1) definitions of $\mathbf{E}$ and $\mathbf{B}$ can be expanded to SU(2) forms:

$$\mathbf{E} = -(\nabla \times \mathbf{A}) - \frac{\partial \mathbf{A}}{\partial t} - \nabla\phi, \quad \mathbf{B} = (\nabla \times \mathbf{A}) - \frac{\partial \mathbf{A}}{\partial t} - \nabla\phi. \tag{4.5}$$

Furthermore, the U(1) Lorentz force law, Eq. (4.3), can hardly apply in this situation because the solenoid is electrically neutral to the test electrons and therefore $E=0$ along the two paths. Using the definition of B in Eq. (4.5), the force law in this SU(2) situation is

$$\mathscr{F} = e\mathbf{E} + ev \times \mathbf{B} = e\left(-(\nabla \times \mathbf{A}) - \frac{\partial \mathbf{A}}{\partial t} - \nabla\phi\right)$$
$$+ ev \times \left((\nabla \times \mathbf{A}) - \frac{\partial \mathbf{A}}{\partial t} - \nabla\phi\right), \tag{4.6}$$

but we should note that Eqs. (4.3) and (4.4) are *still valid* for the conventional theory of electromagnetism based on the U(1) symmetry Maxwell's equations provided in Table 3.1 and associated with the group U(1) algebra. They are *invalid* for the theory based on the modified SU(2) symmetry equations also provided in Table 3.1 and associated with the group SU(2) algebra.

[ooo]Ryder, L.H., *Quantum Field Theory*, 2nd edn. (Cambridge University Press, 1996).

The typical explanation of the Aharonov–Bohm effect continues with the observation that a phase difference, δ, between the two test electrons is caused by the presence of the solenoid:

$$\Delta\delta = \Delta\alpha_1 - \Delta\alpha_2 = \frac{e}{\hbar}\left(\int_{l_2}\mathbf{A}\cdot dl_2 - \int_{l_1}\mathbf{A}\cdot dl_1\right)$$

$$= \frac{e}{\hbar}\int_{l_2-l_1}\nabla\times\mathbf{A}\cdot dS = \frac{e}{\hbar}\int\mathbf{B}\cdot dS = \frac{e}{\hbar}\varphi_M, \qquad (4.7)$$

where $\Delta\alpha_1$ and $\Delta\alpha_2$ are the changes in the wave function for the electrons over paths 1 and 2, S is the surface area, and, φ_M is the *magnetic flux*, defined as

$$\varphi_M = \iint \mathbf{A}_\mu(x)dx^\mu = \iint F_{\mu\nu}d\sigma^{\mu\nu}. \qquad (4.8)$$

Now, we can extend this explanation further, by observing that the local phase change at III of the wave function of a test wave or particle is given by

$$\Phi = \exp\left[ig_m\iint\mathbf{A}_\mu(x)dx^\mu\right] = \exp\left[ig_m\varphi_M\right]. \qquad (4.9)$$

Φ, which is proportional to the magnetic flux, φ_M, is known as the *phase factor* and is *gauge-covariant*. Furthermore, the phase factor Φ measured at position III, is the *holonomy* of the *connection*, $\mathbf{A}_\mu$; and g_m is the SU(2) *magnetic charge density*.

We next observe that φ_M is in units of volt-seconds (V.s) or kg.m^2/A.s^2 = J/A. From Eq. (4.7) it can be seen that $\Delta\delta$ and the phase factor, Φ, are dimensionless. Therefore, we can make the prediction that if the magnetic flux, φ_M, is known and the phase factor, Φ, is measured (as in the Aharonov–Bohm situation), *the magnetic charge density, g_m, can be found through the relation*

$$g_m = \ln(\Phi)/i\varphi_M. \qquad (4.10)$$

Continuing the explanation: as was noted above, $\nabla\times\mathbf{A} = 0$ outside the solenoid and the situation must be redefined in the following way. An electron on path 1 will interact with the $\mathbf{A}$ field oriented in the positive direction. Conversely, an electron on path 2 will interact with the **A** field oriented in the negative direction. Furthermore, the **B** field can be defined with respect to a local stationary component $\mathbf{B}_1$ which is confined to the solenoid and a component $\mathbf{B}_2$ which is either a standing wave or propagates:

$$\begin{aligned}
\mathbf{B} &= \mathbf{B}_1 + \mathbf{B}_2, \\
\mathbf{B}_1 &= \nabla \times \mathbf{A}, \\
\mathbf{B}_2 &= -\frac{\partial \mathbf{A}}{\partial t} - \nabla\phi.
\end{aligned} \tag{4.11}$$

The magnetic flux density, $\mathbf{B}_1$, is the confined component associated with U(1) $\times$ SU(2)symmetry and $\mathbf{B}_2$ is the propagating or standing wave component associated *only* with SU(2) symmetry. In a U(1) symmetry situation, $\mathbf{B}_1$ = components of the field associated with U(1) symmetry, and $\mathbf{B}_2 = 0$.

The electrons traveling on paths 1 and 2 require different times to reach III from X, owing to the different distances and the opposing directions of the potential **A** along the paths l_1 and l_2. Here we address only the effect of the opposing directions of the potential **A**, i.e. the distances traveled are identical over the two paths. The change in the phase difference due to the presence of the **A** potential is then

$$\begin{aligned}
\Delta\delta &= \Delta\alpha_1 - \Delta\alpha_2 \\
&= \frac{e}{\hbar}\left[\int_{l_2}\left(-\frac{\partial \mathbf{A}_+}{\partial t} - \nabla\phi_+\right)\cdot dl_2 - \int_{l_1}\left(-\frac{\partial \mathbf{A}_-}{\partial t} - \nabla\phi_-\right)dl_1\right]\cdot d\mathbf{S} \\
&= \frac{e}{\hbar}\int \mathbf{B}_2 \cdot d\mathbf{S} = \frac{e}{\hbar}\varphi_M.
\end{aligned} \tag{4.12}$$

There is no flux density $\mathbf{B}_1$ in this equation since this equation describes events outside the solenoid, but only the flux density $\mathbf{B}_2$ associated with group SU(2) symmetry; and the $+$ and $-$ indicate the direction of the **A** field encountered by the test electrons — as discussed above.

We note that the phase effect is dependent on $\mathbf{B}_2$ and $\mathbf{B}_1$, but not on $\mathbf{B}_1$ alone. Previous treatments found no convincing argument around the fact that whereas the Aharonov–Bohm effect depends on an interaction with the **A** field outside the solenoid, **B**, defined in U(1) electromagnetism as $\mathbf{B} = \nabla \times \mathbf{A}$, is zero at that point of interaction. However, when **A** is defined in terms associated with an SU(2) situation, that is not the case, as we have seen.

We depart from former treatments in other ways. Commencing with a *correct* observation that the Aharonov–Bohm effect depends on the topology of the experimental situation and that the situation is not simply connected, a former treatment then erroneously seeks an explanation of the effect in the connectedness of the U(1) gauge symmetry of conventional electromagnetism, but for which (1) the potentials are ambiguously defined [the U(1) **A** field is gauge-invariant] and (2) in U(1) symmetry $\nabla \times \mathbf{A} = 0$ outside the solenoid.

Furthermore, whereas a former treatment again makes a *correct* observation that the non-Abelian group, SU(2), is simply connected and that the situation is governed by a multiply connected topology, the author fails to observe that the non–Abelian group SU(2) defined over the integers modulo 2, SU(2)/Z_2, is, in fact, multiply connected. Because of the two paths around the solenoid, it is this group which describes the topology underlying the Aharonov–Bohm effect.[ppp] SU(2)/$Z_2 \approx$ SO(3) is obtained from the group SU(2) by identifying pairs of elements with opposite signs. The $\Delta\delta$ measured at location III in Fig. 4.1 is derived from a *single* path in SO(3)[qqq] because the *two* paths through locations I and II in SU(2) are regarded as a *single* path in SO(3). This path in SU(2)/$Z_2 \approx$ SO(3) cannot be shrunk

---

[ppp] Barrett, T.W., "Electromagnetic phenomena not explained by Maxwell's equations," in Lakhtakia, A. (ed.), *Essays on the Formal Aspects of Maxwell's Theory* (World Scientific, Singapore, 1993) pp. 6–86.

_____, "Sagnac effect.," in: Barrett, T.W. & Grimes, D.M., (eds.), *Advanced Electromagnetism: Foundations, Theory, Applications* (World Scientific, Singapore, 1995) pp. 278–313.

_____, "The toroid antenna as a conditioner of electromagnetic fields into (low energy) gauge fields," *Speculations in Science and Technology*, vol. 21, no. 4, pp. 291–320, 1998.

[qqq] **O(n) Group Algebra**
The orthogonal group O(n), is the group of transformation (including inversion) in an n-dimensional Euclidean space. The elements of O(n) are represented by $n \times n$ real orthogonal matrices with $n(n-1)/2$ real parameters satisfying $AA^t = 1$.

O(3) Group Algebra
The orthogonal group O(3), is the well-known and familiar group of transformations (including inversions) in three-dimensional space with three parameters, these parameters being the rotation or Euler angles (α, β, γ). O(3) leaves the distance squared, $x^2 + y^2 + z^2$, invariant.

to a single point by any continuous deformation and therefore adequately describes the multiply-connectedness of the Aharonov–Bohm situation. Because the former treatment failed to note the multiply-connectedness of the $SU(2)/Z_2$ description of the Aharonov–Bohm situation, it *incorrectly* fell back on a $U(1)$ symmetry description.

Now back to the main point of this excursion to the Aharonov–Bohm effect: the reader will note that the author appealed to topological arguments to support the main points of his argument. Underpinning the $U(1)$ Maxwell theory is an Abelian algebra; underpinning the $SU(2)$ theory is a non-Abelian algebra. The algebras specify the form of the equations of motion. However, whether one or the other algebra can be (validly) used can only be determined by topological considerations.

SO(3) Group Algebra

The collection of matrices in Euclidean three-dimensional space which are orthogonal and moreover for which the determinant is $+1$ is a subgroup of $O(3)$. $SO(3)$ is the special orthogonal group in three variables and defines rotations in three-dimensional space.

Rotation of the Riemann sphere is a rotation in $\mathfrak{R}^3$ or $\xi - \eta - \zeta$ space, for which

$$\xi^2 + \eta^2 + \zeta^2 = 1, \quad \xi = \frac{2x}{|z|^2 + 1}, \quad \eta = \frac{2y}{|z|^2 + 1}, \quad \zeta = \frac{|z|^2 - 1}{|z|^2 + 1}, \quad z = x + iy = \frac{\xi + i\eta}{1 - \zeta},$$

$$U_\xi(\alpha) = \frac{1}{\sqrt{2}} \begin{pmatrix} 1 & -1 \\ 1 & 1 \end{pmatrix} \begin{pmatrix} e^{i\alpha/2} & 0 \\ 0 & e^{-i\alpha/2} \end{pmatrix} \frac{1}{\sqrt{2}} \begin{pmatrix} 1 & 1 \\ -1 & 1 \end{pmatrix} = \begin{pmatrix} \cos\alpha/2 & i\sin\alpha/2 \\ i\sin\alpha/2 & \cos\alpha/2 \end{pmatrix}$$

or $\pm U_\xi(\alpha) \to R_1(\alpha),$

$$U_\eta(\beta) = \frac{1}{\sqrt{2}} \begin{pmatrix} 1 & -i \\ -i & 1 \end{pmatrix} \begin{pmatrix} e^{i\beta/2} & 0 \\ 0 & e^{-i\beta/2} \end{pmatrix} \frac{1}{\sqrt{2}} \begin{pmatrix} 1 & i \\ i & 1 \end{pmatrix} = \begin{pmatrix} \cos\beta/2 & -\sin\beta/2 \\ \sin\beta/2 & \cos\beta/2 \end{pmatrix}$$

or $\pm U_\eta(\beta) \to R_2(\beta),$

$$U_\zeta(\gamma) = \frac{1}{\sqrt{2}} \begin{pmatrix} 1 & 0 \\ 0 & 1 \end{pmatrix} \begin{pmatrix} e^{i\gamma/2} & 0 \\ 0 & e^{-i\gamma/2} \end{pmatrix} \frac{1}{\sqrt{2}} \begin{pmatrix} 1 & 0 \\ 0 & 1 \end{pmatrix} = \begin{pmatrix} \cos\gamma/2 & -\sin\gamma/2 \\ \sin\gamma/2 & \cos\gamma/2 \end{pmatrix}$$

or $\pm U_\zeta(\gamma) \to R_3(\gamma).$

which are mappings from $SL(2,C)$ to $SO(3)$. However, as the $SL(2,C)$ are all unitary with determinant equal to $+1$, they are of the $SU(2)$ group. Therefore $SU(2)$ is the covering group of $SO(3)$. Furthermore, $SU(2)$ is simply connected and $SO(3)$ is multiply connected.

A simplification of the above is

$$U_\xi(\alpha) = e^{i(\alpha/2)\sigma_1}, \quad U_\eta(\beta) = e^{-i(\beta/2)\sigma_2}, \quad U_\zeta(\gamma) = e^{i(\gamma/2)\sigma_3},$$

$$\text{where} \quad \sigma_1 = \begin{pmatrix} 0 & 1 \\ 1 & 0 \end{pmatrix}, \quad \sigma_2 = \begin{pmatrix} 0 & -i \\ i & 0 \end{pmatrix}, \quad \sigma_3 = \begin{pmatrix} 1 & 0 \\ 0 & -1 \end{pmatrix}.$$

$\sigma_1, \sigma_2, \sigma_3$ are the Pauli matrices.

5. Discussion

We have shown the fundamental explanatory nature of the topological description of solitons, instantons and the Aharonov–Bohm effect — and hence electromagnetism. In the case of electromagnetism, we have shown in previous chapters that, given a Yang–Mills description, electromagnetism can and should be extended, in accordance with the topology with which the electromagnetic fields are associated.

This approach has further implications. If the conventional theory of electromagnetism, i.e. "Maxwell's theory," which is of U(1) symmetry form, is but the simplest *local* theory of electromagnetism, then those pursuing a unified field theory may wish to consider as a candidate for that unification, not only the simple local theory, but other electromagnetic fields of group symmetry higher than U(1). Other such forms include symplectic gauge fields of higher group symmetry, e.g. SU(2) and above.

Index

.